GEN

FORT WAYNE
IND.
EIGHTH
ANNUAL REPORT OF
BOARD OF PARK COMMISSIONERS
1912

THE Cover design of this report is from a painting by Ralph Dille, and is a scene on the banks of the St. Mary's along Swinney Park. The tree is an American Sycamore, 17 feet and 8 inches in circumference.

Frank J Schlebecker
1210 Spy Run Ave
City

EIGHTH ANNUAL REPORT
BOARD of PARK COMMISSIONERS

1912

FORT WAYNE, INDIANA

PLATES BY
FORT WAYNE ENGRAVING COMPANY

PRINTED BY
SINGMASTER PRINTING COMPANY
FORT WAYNE, INDIANA

Donors of Foster Park.

BOARD OF PARK COMMISSIONERS

1905-1912.

Pursuant to an Act passed by the Legislature and approved March 6th, 1905, the Department of Public Parks was established and the Commissioners were appointed to serve, respectively, one, two, three and four years.

Name—	1st Term.	2nd Term.	3rd Term.
David N. Foster........	1905-1907	1907-1911	1911-1915
*Oscar W. Tresselt......	1905-1906	1906-1910	1910-1914
†Ferdinand Meier........	1905-1909	1909-1913	
‡Joseph M. Singmaster...	1905-1908	1908-1912	
Louis W. Dorn.........	1912-1913	1913-1917	
Louis Fox	1912-1916		
E. F. Yarnelle..........	1912-1914		

*Resigned September 9th, 1912, E. F. Yarnelle appointed to fill unexpired term.

†Resigned February 1st, 1912, Louis W. Dorn appointed to fill unexpired term.

‡Term expired, Louis Fox appointed as his successor.

David N. Foster, President.
Louis Fox, Vice President.
Louis W. Dorn.
E. F. Yarnelle.

George E. Kessler.....................Landscape Architect
Marriott Price.................................Engineer
A. W. Goers..............................Superintendent
Lillian C. Busch................Chief Bureau of Assessment
Carl J. Getz...................................Forester
Charles J. Steiss..............................Secretary

Entrance to Swinney Park, Fort Wayne, Indiana.

REPORT OF THE BOARD OF PARK
COMMISSIONERS FOR THE YEAR 1912
RALPH DILLE

Fort Wayne, Indiana, December 31, 1912.

Hon. Jesse Grice, Mayor:

DEAR SIR: Our Board has now operated for something more than a year and a half under the Park law enacted by the legislature of 1911 for Indiana cities of the first and second class. However, no work was attempted under this law until 1912. Early in that year at the request of our Board the City Council divided the city into four park districts bounded generally speaking, by Calhoun Street North and South, and by the Pennsylvania and Wabash Railroads East and West.

PLANS OF GEORGE E. KESSLER.

There was at this time pending before our community and the City Council a proposition to follow out the suggestions and the plans of Mr. George E. Kessler, the celebrated Landscape Architect, for the purchase of our entire river front, and many adjacent park strips of inexpensive lands, by the issuance of city bonds amounting to $200,000. Pending the settlement of this question it did not seem wise to undertake any very considerable work of a general nature.

The discussion ran on through the spring months, and it was not until summer that the Council finally voted down the proposition. It was generally admitted that a vast preponderance of those who pay eighty if not ninety per cent of the taxes of the city favored the measure, but it was argued that in so large an enterprise, as was here contemplated, the Council should not be asked to assume so great a responsibility without a direct vote of the people in its favor. On investigation it was found the law provided no method of submitting the question to the voters and it therefore became necessary to await the meeting of the legislature of 1913 before taking further action.

INITIATIVE AND REFERENDUM.

Our Board will then have prepared and introduced into that body an amendment to the Cities and Towns Act of 1905, giving the City Council the right to submit to the people the question of the issuance of bonds for the purchase of park grounds, the building of a garbage plant, or for an auditorium, or for the extension of the water works system, or other needed public utilities. It will provide for both the initiative and referendum upon questions of that character, but will require a large percentage of the voters to petition for such action, in order that trifling questions shall not make necessary too frequent elections. This would seem to be much-needed legislation and it is confidently expected the bill will become a law.

HEAVY PENALTY.

Fort Wayne cannot too quickly secure the lands necessary to complete a park system for a city of 250,000 inhabitants, for land values are constantly increasing and a heavy penalty will have to be paid for each year's delay. It is likely that our children now living will live to see Fort Wayne a city of that population, and while there would be neither sense nor justice in putting upon this generation the whole burden of both purchasing and improving these properties, the interests of our children and those who shall come after us, imperatively demand that prompt action be taken along the line suggested.

INCREASE IN VALUES.

Under your predecessor, and just at the close of his administration in 1908, after much persuasion on the part of the members of our Board, the then City Controller and City Council consented to buy the fifteen acres of forest, constituting Weisser Park, for the sum of $10,500. In less than four years that piece of ground has so grown in value that it would now readily bring over double the purchase price. In other words, Fort Wayne could get all her money back by selling less than half of it, and own the remainder without its costing her a cent. She could sell one-quarter of Reservoir Park and get back all she paid for the whole of it, and have three-quarters of it left. Facts such as these should have great weight with our people and with our city authorities when propositions are made for the acquirement of other needed park properties.

SPECIAL ASSESSMENTS.

Early in the spring of 1912 our Board realized that it would take years in which to complete Lakeside Park if it had to be done out of our regular revenues. That community had always shown so much of public spirit that the Board confidently proposed to it that a special park district be made of that portion of the city, and that a special tax, averaging about $8.00 per lot should be levied upon all property within the lines fixed to raise the necessary funds to at once finish the work. A meeting of the interested citizens was called, a large attendance was had, the plan was explained and they voted without a dissenting voice to have it done. The cost was somewhat more than was estimated, occasioned by adding somewhat to the amount of work to be done, and by the advance in the price of labor and of teams, but at a total expense of $17,500 the work has been completed, and "Lakeside" and "Forest Park" and other smaller adjacent additions, now have a park that is not second in beauty to any in the city. The effect of the completion of this work has been to add immensely to the property values of that district, again demonstrating the fact, that moneys spent by the city in purchasing park properties and in improving them are quickly returned to the treasury in the added taxable values of adjacent real estate, and this stream of increased city revenues is a continuing one for all the years to come.

BOULEVARDS.

Owners of property adjacent to the proposed boulevards are insistent in their demands upon our Board that work upon these Boulevards be not longer delayed. Already therefore we have let the contract for the grading and partial improvement of the section of Rudisill Boulevard from Hanna Street East

Winter Scene in Weisser Park, Fort Wayne, Indiana.

to the first alley West of Thomas Street. All the ground necessary to make a forty-foot roadway and the thirty-foot parkway on either side, one hundred feet in all, was dedicated to the city without charge by the owners of the Grand Boulevard Addition. Other sections of this Boulevard will be let early in the coming year, followed by work on sections of Anthony and State Boulevards.

DEMANDS FOR MORE PARKS.

In our report to your honor a year ago we said: "If Fort Wayne continues to grow as rapidly as she is now doing, it will not be many years before the health and comfort of her citizens and especially of her industrial population, will imperatively demand four additional large parks of at least 100 acres each and well supplied with large forest trees, affording abundant shade, without which park areas are of little use; one of these to be located in the Eighth Ward; another in the Southeast section of the city; a third in the Northwest, and a fourth along the bank of the St. Mary's, commencing just South of the Broadway pumping station." We further suggested that "here are four magnificent opportunities for four or more wealthy citizens of Fort Wayne to immortalize their names," by the purchase and gift to the city of one of these much needed parks.

THE DAVID N. AND SAMUEL M. FOSTER PARK.

Our suggestion has already borne fruit, for during the year two citizens of Fort Wayne, Col. D. N. Foster, a member of this Board, and his brother, Samuel M. Foster, have purchased and given to the city for park purposes a most beautiful strip of land bordering the bank of the St. Mary's, extending from the Broadway pumping station to a point about two miles distant, which is shortly to be extended by the same donors until the park shall reach the Stellhorn bridge, a total distance of about four miles. The ground varies in width from 1,000 to 2,000 feet, is much of it heavily wooded, and particularly adapted for park purposes, and will contain about 110 acres. It will be legally known as "The David N. and Samuel M. Foster Park," but will be commonly designated as "Foster Park." Some inexpensive improvements were made upon the

"Honeymoon Path," Foster Park, Fort Wayne, Ind.

grounds during the past year, and they were largely used by the general public, and were particularly popular with the children.

DEDICATION.

The Park was formally dedicated July 4th, on which day it was estimated the grounds were visited by at least 25,000 people. It is hoped to let the contract the coming year for the permanent improvement of this park, which will include a low dam in the river near the pumping station, giving a boating stage of water of about three and a half miles in length; a boat house near the dam to accommodate boaters in the summer, and skaters in the winter; a Pavilion on the high ground just outside the wooded portion, about 1500 feet from the park entrance, which will be rented as a place for refreshments, but will also afford temporary shelter in case of a sudden shower; extensive tennis grounds and an athletic field farther to the South, with a shelter house between them; the construction of a roadway placed back against the private property, to be used as a street or parkway by abutting lot owners, and the laying out of the necessary walks. As the Southwest park district has secured its park without cost it is believed the residents of that section of the city will gladly assist in its speedy improvement. The Foster brothers desire that special mention be made of the generosity and public spirit of Judge Wm. J. Vesey, who, when approached by them for the purchase of several acres of his land needed for the proposed park and was informed of the use to which it was to be put, refused to accept any compensation for it, and deeded it without charge.

THE JOHN H. VESEY PARK.

During the year just closed other citizens have shown the same spirit as that exhibited by the Foster brothers, which has led to the dedication to the public of other smaller but pretty parks. One of the most beautiful of these was the gift of Irvington Park, North of the city, through which Spy Run threads its way, by the late John H. Vesey. Irvington Park consisted of only that part of the grounds North of Spy Run. After his death his widow very generously fixed a particularly low price on the portion of the grounds South of the Run,

Winter Scene Lakeside Park, Showing a Section of the Lagoons.

Boulevard. This boulevard strip before its donation had been all graded and curbed and planted with $3,000 worth of shrubbery by the company giving it.

Winter Scene Lakeside Park, Showing a Section of the Lagoons.

[illegible] of Lakeside Park and Entrance to [illegible] Park [illegible].

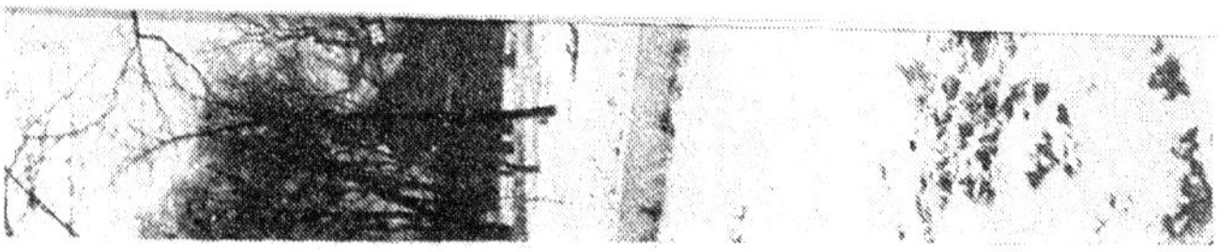

for the purpose of enabling the city to add them to those already dedicated, thus doubling the size of the park. Our Board did not have the available funds necessary to make the purchase but that the ground might not be lost it was purchased for the sum of $5,400 by Mr. Louis Fox, a member of this Board, who will hold it in trust for the city until funds are available for its acquirement. In honor to the memory of Mr. Vesey, the name of the park has been changed from Irvington Park to "The John H. Vesey Park," and will be popularly known as "Vesey Park."

PONTIAC PLACE PARK.

The gentlemen who acquired the Concordia College Farm in the Southeast section of the city and have laid out Pontiac Place Addition have dedicated to the city for park purposes a strip of ground 1,264 feet long lying between old Pontiac Street and the new Pontiac Boulevard, which when fully improved will make a very attractive small neighborhood park. This company has also dedicated without charge the ground upon the East side of Anthony Boulevard necessary to make it 100 feet wide, and has at its own expense graded the 30 foot park strip and laid the 6 foot cement sidewalk along the whole length of the West line of Pontiac Place Addition.

HIRONS' PARK.

In the recently laid out Hirons' Addition South of the city on the Piqua Road its owner, Mr. Albert R. Hirons, dedicated an attractive three-cornered piece of ground to the city, which will be known as Hirons Park, in honor of the gentleman named. It has already been improved, and trees and shrubbery planted.

ADDITIONS TO LAKESIDE PARK.

Before the commencement of the improvements to Lakeside Park last spring it was further enlarged by the donation of the Boulevard Realty Company of Lot A, and the half-mile strip of ground between the two driveways on Forest Park Boulevard. This boulevard strip before its donation had been all graded and curbed and planted with $3,000 worth of shrubbery by the company giving it.

Guldin Playgrounds, Dedication Exercises, May 20th, 1911.

The Parham warehouse which until recently stood within the lines of Lakeside Park, was purchased by the Board, with the land about it, during the year for the sum of $2,800. The warehouse was carefully taken down and the lumber removed to the city lots on Nelson and Jones Streets, and used for the erection of a large barn, 46x120 feet, across the rear of said lots, which will be utilized by our Board for the purpose of housing our machinery, park seats, spraying wagons, etc., during the winter months. In one corner of this barn there has been constructed a cold storage cellar 16x26 feet in dimensions and 6 feet in depth, for the use of the City Forester. Trees received from nurseries will be here preserved in perfect condition until time for planting. The front portion of the lots will be made into a miniature park.

CAMP ALLEN.

Through the aid of Mr. Howell C. Rockhill the Board purchased early last spring about three acres of land lying along the West bank of the St. Mary's, opposite Thieme Drive, at a cost of $1,284.83. Other land adjacent to this piece should be secured for the double purpose of affording a neighborhood park to that locality and to make a prettier setting for the entrance to Swinney Park at Washington Boulevard West.

THIEME DRIVE.

A very large amount of filling has been done upon Thieme Drive during the past year, without expense to the benefited property, and it is nearing completion. A local district assessment will shortly be made to acquire the property necessary to widen the Drive from forty to sixty-five feet.

BROADWAY PARKWAY.

The city has also come into the possession during the past year of two parcels of ground along the St. Mary's river bank lying South of the approach to the Bluffton Road bridge and West of Broadway, by donation of Montgomery G. Beaver and wife, Frank M. Miller and wife, Florence E. Nicholson and husband, Catherine Kimball, R. L. Romy and wife and Samuel S. Fisher and wife.

Guldlin Playgrounds, Girls' Section.

SWIMMING POOL AT SWINNEY PARK.

During the coming year the Board is hopeful that it may be able to put in a capacious concrete swimming pool in Swinney Park at the North end of the grove, supplied with pure water, changed frequently. Located at that point it will be removed from the grounds used for picnic purposes, and can be successfully screened from view by a proper arrangement of the clothes lockers.

FORESTRY DEPARTMENT.

Under the energetic management of our forester, Mr. Carl J. Getz, a graduate of the Forestry Department of Purdue University, much valuable work has been accomplished in this department. It has been to a considerable extent self-supporting and is likely to be more so in the future, as the working force becomes better organized and therefore more efficient. The department has been able to do only a portion of the private work that has been demanded of it. Attention is called to the full report of Forester Getz appended hereto.

CHILDREN'S PLAYGROUNDS.

In our report of one year ago we expressed the conviction that the Trustees of the School City of Fort Wayne should, as soon as possible, take charge of the city's playgrounds and thus relieve them from the uncertain support to be derived from the generosity of private citizens or from the restricted revenues of the Park Board. These public playgrounds have come to be recognized as a necessary adjunct to our public schools. Our Board is gratified to receive the information that a bill is being prepared and will be introduced into the legislature and is likely to become a law enabling the School Trustees to take over this work, and providing a suitable revenue for its support. The relief thus afforded our revenues will enable us to install many play devices in our parks which can be safely used by children without the supervision required upon the playground, thus greatly adding to the attractiveness of our parks to both parents and children.

HOW VIEWS DO VARY.

It is sometimes a bit amusing to see what different views

A Beautiful Scene on the Maumee at "Shady Nook Point."

different people have of what should be the chief attraction of our parks. To perhaps the largest number of people our floral display is the greatest attraction and they would have the Board to constantly enlarge it. Some time since we had a prominent lady living near Reservoir Park wait upon us in a very indignant frame of mind complaining that the floral display at Reservoir Park did not equal that at McCulloch Park.

Next comes along a man who cards the paper with a snarl at the Board for having removed the old eye-sore of a bath house at Lakeside Park and for not spending its revenues to remove hundreds of tons of snow from the ice in our parks after a ten-inch snow-fall, by the use of men and teams, instead of entrusting a portion of this work to the Boy Scouts, who have generously volunteered to do it without cost. This man has no eye for parks in which he cannot bathe or lakes upon which he cannot skate, and the beautiful floral display he dismisses with contempt. Others again find in boating the chief attraction; hundreds enjoy tennis, the boys baseball; the girls the play apparatus and the more quiet sports. To others the great delight is the family or society picnic or evening lunch under the great trees.

Some men can see little advantage in neighborhood parks, so popular with the mother and her children who has to count her street car fares, and so they would stop providing these smaller areas and would go to the outskirts and buy one great big park with miles of automobiling, horseback riding, etc. Still others complain that we have not long since commenced the establishment of an extensive zoological garden, always a great park attraction. The fact is that all these facilities for recreation and for sport should be afforded our people and will be as rapidly as the revenues of this Board allow.

CUT IN REVENUE.

In this connection it is pertinent to refer to the one cent cut in our levy for the year 1913. While this is unfortunate and will prevent some much needed work the Board does not want to be understood as complaining of the action of the City Council. No Council has ever dealt more liberally by this Board than the present one and we confidently expect it will

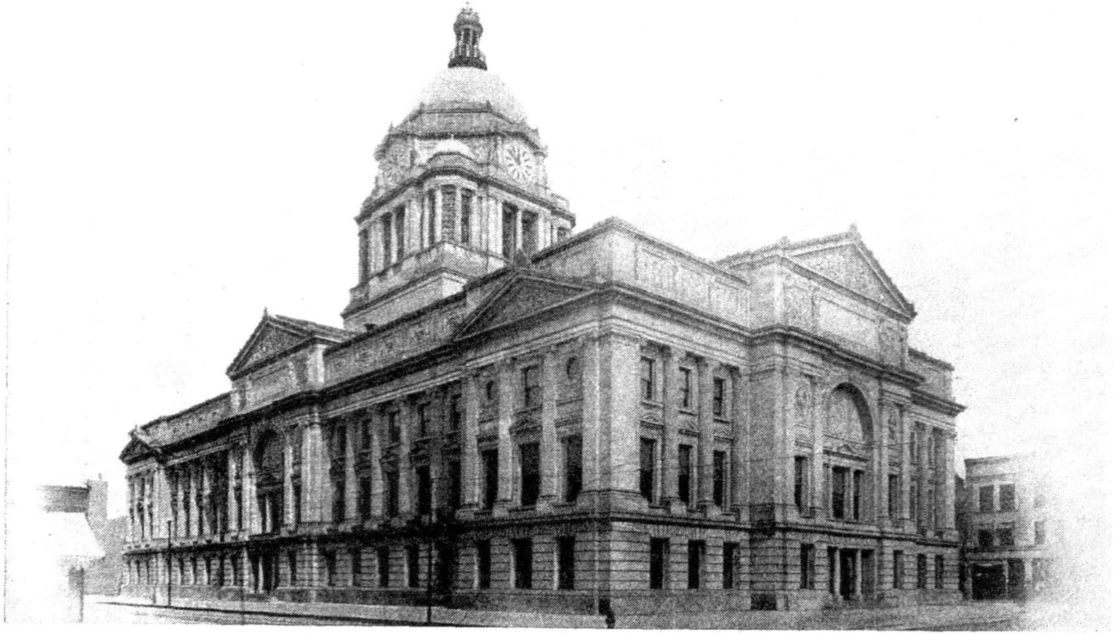

Allen County Court House, Fort Wayne, Indiana.

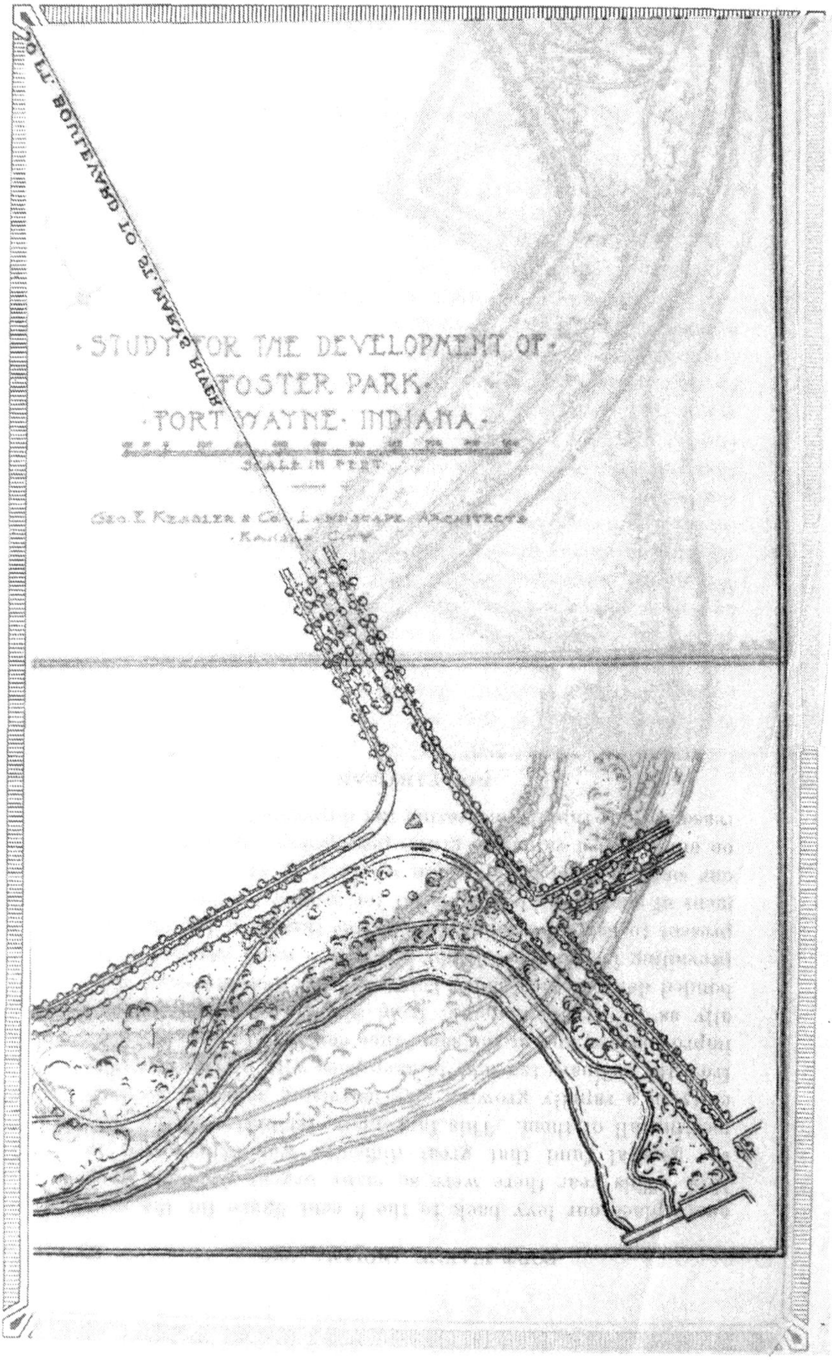

TO LT. BOULEVARD TO ST. MARYS RIVER
·STUDY FOR THE DEVELOPMENT OF·
·FOSTER PARK·
·FORT WAYNE· INDIANA·
SCALE IN FEET
GEO. E. KESSLER & CO. LANDSCAPE ARCHITECTS
·KANSAS CITY·

Allen County Court House, Fort Wayne, Indiana.

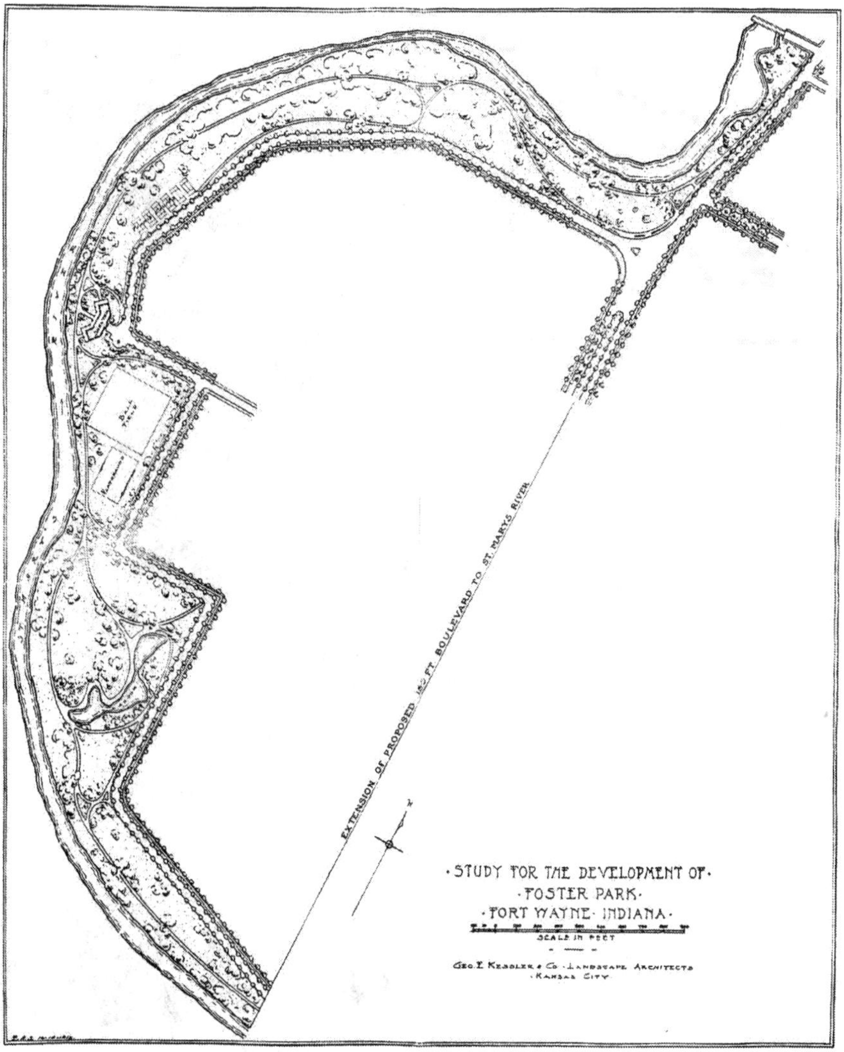
EXTENSION OF PROPOSED 150 FT. BOULEVARD TO ST. MARYS RIVER
·STUDY FOR THE DEVELOPMENT OF·
·FOSTER PARK·
·FORT WAYNE· INDIANA·
SCALE IN FEET
GEO. E. KESSLER & CO. · LANDSCAPE ARCHITECTS
· KANSAS CITY

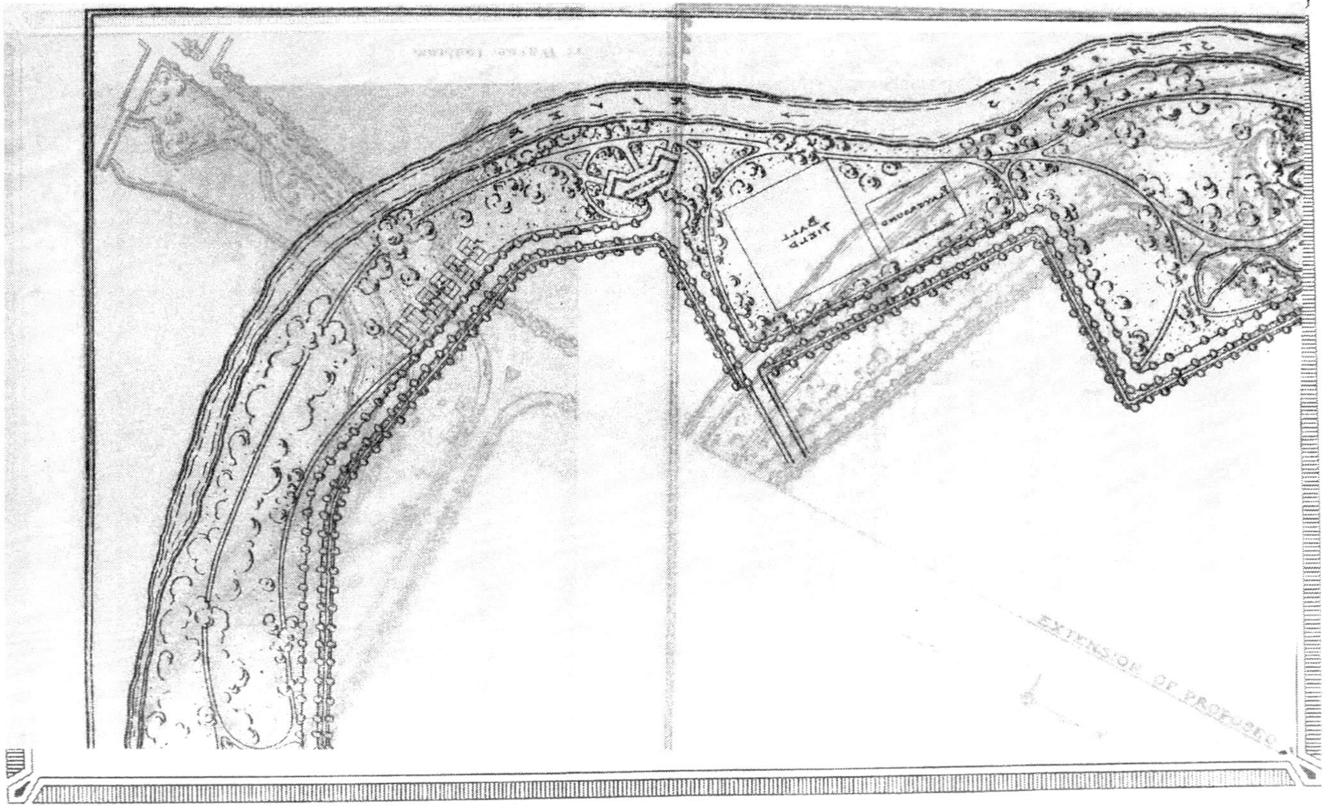
BALL FIELD
PLAYGROUND
EXTENSION OF PROPOSED

again place our levy back to the 9 cent figure for the year 1914. This year there were so many urgent demands upon the general fund that great difficulty was experienced in meeting all of them. This fact again demonstrates the difficulty in a rapidly growing city in raising sufficient revenue from the ordinary tax levy to keep pace with necessary public improvements, and at the same time continue to pay off annually as we are now doing, from $15,000 to $20,000 of our bonded debt, already much below the low constitutional limit prevailing in Indiana. Might it not be a wiser course for the present to use our sinking fund, after this year, for the payment of short time bonds issued for needed purposes and let our small bonded debt remain where it is, as its percentage on our assessed valuation grows perceptibly less each year by reason of our rapidly increasing tax duplicate?

NON-PARTISAN.

Under the law of 1905, as well as that of 1911, our Board is made to consist of four members to be appointed by the mayor, not more than two from any one political party. The result has been that the question of whether the country should have a high tariff or a low tariff or no tariff at all has had no weight in the conduct of our business for the city. When this make-up of our Board was first proposed it was objected to on the ground that it would constantly be evenly divided and thus be unable to do business, and so to avoid that expected-to-be very frequent situation it was provided that your honor should be called in to cast the deciding vote. During the eight years this Board has been in existence it has never been necessary to avail ourselves of that provision of the law. But this has not resulted in depriving us of your frequent presence at our meetings, and of having the benefit of your advice and hearty co-operation in our work.

We desire also to make acknowledgment of the many courtesies at the hands of the City Council and of the City Controller and to bear testimony to the promptness and efficiency with which this Board has been assisted in legal matters by City Attorney Hogan.

EMPLOYEES.

Our office force with greater experience has grown in

Entrance to Lawton Park, Soldiers Monument in Foreground.

efficiency and has never hesitated when necessary to put in much overtime work. The first large work of our assessment bureau was in getting out our rolls for the improvement of Lakeside Park, in which nearly two thousand pieces of property were involved. That the assessments were so equitably distributed as to excite practically no complaint is to the credit of Miss Lillian C. Busch, the Chief of that department.

CITIES GET WHAT THEY BID FOR.

Cities, as a rule, get what they bid for. If by good streets, pretty, well kept homes, adequate sewerage, attractive public buildings, plenty of schools, churches, libraries and theaters, and an extensive system of well distributed parks they bid for growth, and for growth of the right character, it is sure to come, and this Board wants no truer illustration of the truth of that statement than is afforded by the rapid growth of this city of Fort Wayne during the past five or six years. If in some small measure the work of our Board has contributed to this rapid growth, we are content.

Respectfully submitted,

DAVID N. FOSTER,
LOUIS FOX,
LOUIS DORN,
E. F. YARNELLE,
Board of Park Commissioners.

St. Mary's River—Foster Park—Looking Down Stream.

SUPPLEMENTARY

THE Park Commissioners whose names are attached hereto, desire to emphasize more fully than is done in the regular report the gratitude which our citizens owe to Col. D. N. Foster and Hon. S. M. Foster for the magnificent gift of "Foster Park," which they have donated to the City of Fort Wayne, and which is to be forever maintained as a public park for all the people. Looking forward into the next fifty years every one who is able to see, and who is filled with confidence in, the progressive spirit of Fort Wayne will realize that the city is bound to grow Southward until it reaches the Stellhorn bridge and that this park is destined to become a great blessing to our citizens. Its donors have acted wisely in giving it while they live, and we hereby express the hope of all our people that the two distinguished gentlemen be spared to this community for many years, and thus may have the satisfaction of seeing these grounds used continually and enjoyed by the thousands who will frequent them. Such generous gifts to our community are too rare to pass them by without particular mention of the splendid civic patriotism which inspire them and without an expression on our part of the hope that other wealthy citizens of Fort Wayne may also be led to build unto themselves monuments equally as noble and as enduring as have our two public-spirited citizens, Col. David N. Foster and Hon Samuel M. Foster—monuments that shall stand for many centuries to come, and grow in beauty and in usefulness as the years go by.

It is not merely egotism, or the love of praise and fame, that prompts noble men and women to make sacrifices and to confer lasting benefits upon their fellow citizens. Such actions usually spring from the consciousness of a higher duty towards

Resolutions By The Board Of Parks Commissioners Accepting The "Foster Parks"

Adopted March 23, 1912.

Whereas our distinguished fellow citizens Col. David N. Foster and Hon. Samuel M. Foster have donated to the city of Fort Wayne, Indiana, certain lands which are fully described in the deed accompanying their letter of donation; and

Whereas these lands, according to the will of the donors, are to be used for Park and recreation purposes subject to the rules and regulations of the authorities to whom the parks of the city are entrusted; and

Whereas the magnanimous donors have hampered the city by no conditions or restrictions excepting those which are determined by the nature and purpose of the splendid gift:

Therefore be it resolved by the Board of Parks Commissioners through the three members whose signatures are attached to this paper. That, acting on behalf and in the name of the City of Fort Wayne, we accept cheerfully and with deeply felt gratitude this gift for the purpose indicated by the grantors and that we hereby officially tender them the heartiest thanks of our entire community which justly deems itself honored highly by such munificence of two of its own members; and be it further resolved:

That we express our appreciation of the magnificent donation by giving to the lands thus conveyed to the city the official designation: "The David N. and Samuel M. Foster Parks", or more briefly "The Foster Parks"; and be it further resolved

That we commend the donors for their public spirited action, through which they have erected unto themselves a grand monument that will, for ages to come, perpetuate not only their names, but the memory of their civic consciousness, of their disinterested love towards their fellow citizens, and of their municipal patriotism, serving as a glorious example to be imitated by the present and by future generations, - a living and growing monument, which with its green boughs, its shaded walks and driveways, its singing birds, its rippling waters, its brilliant flowers and other charms of nature, giving rest and recreation, joy and happiness to thousands of men, women and children, will be more enduring and beneficial by far than a mere record engraved on cold dead stone could ever be, - a monument, that will be the pride of Fort Wayne and a constant delight to every inhabitant of our city as well as to the stranger whom we are glad to entertain as our guest.

Board of Parks Commissioners.

Louis Fox

O. W. Tresselt

Louis Dorn

humanity and stand forth as brilliant beacon lights of disinterested patriotism.

Our city is under obligations to the Foster brothers for having taught this lesson in an effective manner, which is of immeasurable value to the progress of civic conditions, and we feel that it is our duty to give these cheerful givers due credit for setting such a splendid example, worthy of emulation.

In connection herewith we publish a photographic copy of the resolutions of thanks adopted by us at the time of the presentation to the city of the deeds to the property, which resolutions were handsomely engrossed and framed, and a copy presented to each of the donors.

LOUIS FOX.
LOUIS DORN.
E. F. YARNELLE.

View of Thieme Drive Improvement from the West Side of the St. Mary's.

REPORT OF GEORGE E. KESSLER
LANDSCAPE ARCHITECT.

St. Louis, Mo., March 10, 1913.

To the Honorable Board of Park Commissioners, Fort Wayne, Indiana:

GENTLEMEN:—In your report of 1911 there is a recommendation on my part for the acquisition and development of a comprehensive scheme of park and boulevard improvements. In your report to the Mayor, of the same issue, you say with reference to my plans, "He has not planned for a stand-still city."

It is true that many conditions arise which prevent action looking toward securing at least approximately the properties then recommended. It is true also that communities are frequently fearful of debt burdens, and consequently slow in undertaking great projects. Since the issue of this report and its recommendations to your public, you have indeed made certain progress; however, it seems to me an unfortunate condition that your main progress should have been made almost exclusively through the generosity of two of your citizens, namely the acquisition of the properties popularly known as Foster Park, by gift from the Foster brothers. Of course the generosity of these gentlemen stands out all the more strongly in giving you these properties where the city itself is unwilling to make proper provision for its public, yet surely your community is unwilling to stand still. Fortunately you are at the same time operating toward the improvement along Rudisill Boulevard and Anthony Boulevard as well. In the meantime, also, a very material improvement in good appearance has been made on Lakeside Park. On public account, however, it would seem that further acquisi-

View of the St. Mary's and Foster Park, near Entrance.

tions should at the earliest possible time be made, even if these properties can only be leased subject to purchase.

In connecting the Riverside properties if you could only continue the parkway along the St. Mary's River from Foster Park to Swinney, you would make a beginning toward the accomplishment of a really fine park system and beautiful setting for the whole city.

In the Foster Park property you have at once an opportunity for a most excellent boating scheme and a border boulevard, which will immediately attract to itself a residential section that should become a most excellent one.

On the land between the border boulevard and the river is ample room for athletic fields for all of the out-door sports. There is also a very finely timbered section that even today, in natural condition, presents an exceedingly attractive landscape picture. I do not know of any other one property which would deserve, so much as this, immediate attention and a very considerable improvement.

In connection with the Rudisill Boulevard, which ties directly into this property, there will be established a district that will quickly compel other residence districts of the city to undertake similarly worthy improvements.

At the same time there should be no relaxation in the effort to establish the several playgrounds necessary to care for the children in especially the closely built up sections. Some of this can be accomplished on existing property, and necessarily this must go forward coincident with the greater park and boulevard projects. You cannot very long rely upon private endeavor nor private generosity to supply the public needs, but must very soon establish the comprehensive scheme of children's playgrounds, together with all that this means in proper supervision and proper attention to govern the play of the children using these.

Yours very truly,

GEO. E. KESSLER.

Market House and City Hall, Fort Wayne, Indiana.

REPORT OF SUPERINTENDENT FOR 1912.

Fort Wayne, Indiana, December, 31st, 1912.

To the Board of Park Commissioners:

GENTLEMEN:—Herewith is respectfully submitted the report of the Park Department for the year 1912, especially touching upon the work of park maintenance, improvements and betterments. While the principal energies of the department were devoted to new work, at the same time the work of maintenance and improvement was kept on a satisfactory basis. Our parks were decorated with flowers as usual, and these were of an exceptionally good quality. Several specimens of our cannas could not be excelled. The department completed all the work ordered, except the building of the two small bridges, one in Swinney Park, the other at Vesey Park. The construction of these bridges will be taken up the coming spring.

SWINNEY PARK.

The principal work done in Swinney Park was the re-construction of the drive. After re-crowning the roadway, a coat of liquid asphalt was applied; this was covered with a thin layer of screenings, and rolled. I would recommend a dressing of road oil to complete the surface. The roadway leading into the park from Swinney Park Place was treated with road oil, and is in fine condition. One new tennis court was constructed, making four courts now in this park. Backstops were constructed at each of the courts, and also at the baseball diamonds. Most of the trees and shrubs were pruned during the year, which has greatly enhanced their beauty. I would recommend the installation of a new motor pump for the well.

Carpet Bed—General Henry W. Lawton Park.

LAWTON PARK.

The beauty of this park has been greatly enhanced by a number of private and public improvements made along and opposite the western portion of this park. The junk yard at the corner of Fourth and Clinton Streets has been removed, upon which ground the Lake Shore & Michigan Southern Railroad will construct a handsome freight terminal. The paving of Clinton Street, and the construction of a cement walk at a cost of $502.50 along the entire length of the western part of the park, has given the park an entirely different setting. A tennis court was constructed in the eastern portion of the park, and a number of other improvements were made. A fill of about four hundred yards of earth will be made the coming year. This fill is necessary on account of the change in the grade of Clinton Street.

LAKESIDE PARK.

Upon the completion of the grading and excavating for the lagoons which work was done by The Moellering Construction Company, four hundred trees were planted, and all the other trees were pruned. As soon as the weather will permit, several thousand shrubs will be planted and grass seed sown, and tennis courts and basketball banks will be constructed the coming year. When completed this park will be one of the most beautiful in our city. The erection of a tool house and comfort station is recommended. It is intended to fill the lagoons with water, thereby affording boating in the summer and skating in the winter. The floral display in this park was greatly admired during the past year.

FOSTER PARK.

After the acceptance by the Park Commissioners of Foster Park, the department cleared a path about two miles in length through this park, along the banks of the St. Mary's River. Two temporary bridges, a tool house and a comfort station were erected. A fence consisting of two-inch iron posts, painted white, with a substantial wire cloth, was constructed, dividing the private property from the park. Mr. William Breuer very generously purchased a large number of swings, and had them placed in this park. The Wildwood Builders Company constructed a speaker's stand for the dedication exercises on July 4th, and afterwards donated the same to the Park Board, to be utilized as a band stand. Several benches and tables for the use of picnicers were also constructed in this park by this department.

HAYDEN PARK.

The old gravel walks in this park were torn out and replaced with new walks prepared with stone screenings mixed with liquid asphalt and rolled. The park is in a densely populated section, and in close proximity to a school, hence it is much traversed. Many people have acquired the unfortunate habit of walking on the grass along the edges of the walks. Several hundred shrubs will be planted in this park next spring.

WEISSER PARK.

Early last spring we removed a large tree which stood within the baseball diamond. A tennis court was constructed, and a number of benches and tables erected. Some earth was secured when Weisser Park Avenue was paved, to make a fill in the park along Hanna Street. Weisser Park Avenue is now paved up to the park. A comfort station ought to be constructed in this park.

RESERVOIR PARK.

A large number of shrubs were set out in this park during the past season, but many more will be required before this work is completed. Reservoir Park, during the winter months, is used more than any other park in the city, on account of the large number of people going to this park to enjoy skating, and hundreds of children use the hills for coasting.

OTHER PARKS.

The trees and shrubbery in Williams Park are growing nicely, several hundred new shrubs being planted there last fall. Hirons Park, located on Piqua Avenue, was laid out and planted according to the plans drawn by Mr. W. H. Hillary. A large amount of grading was done in Pontiac Place

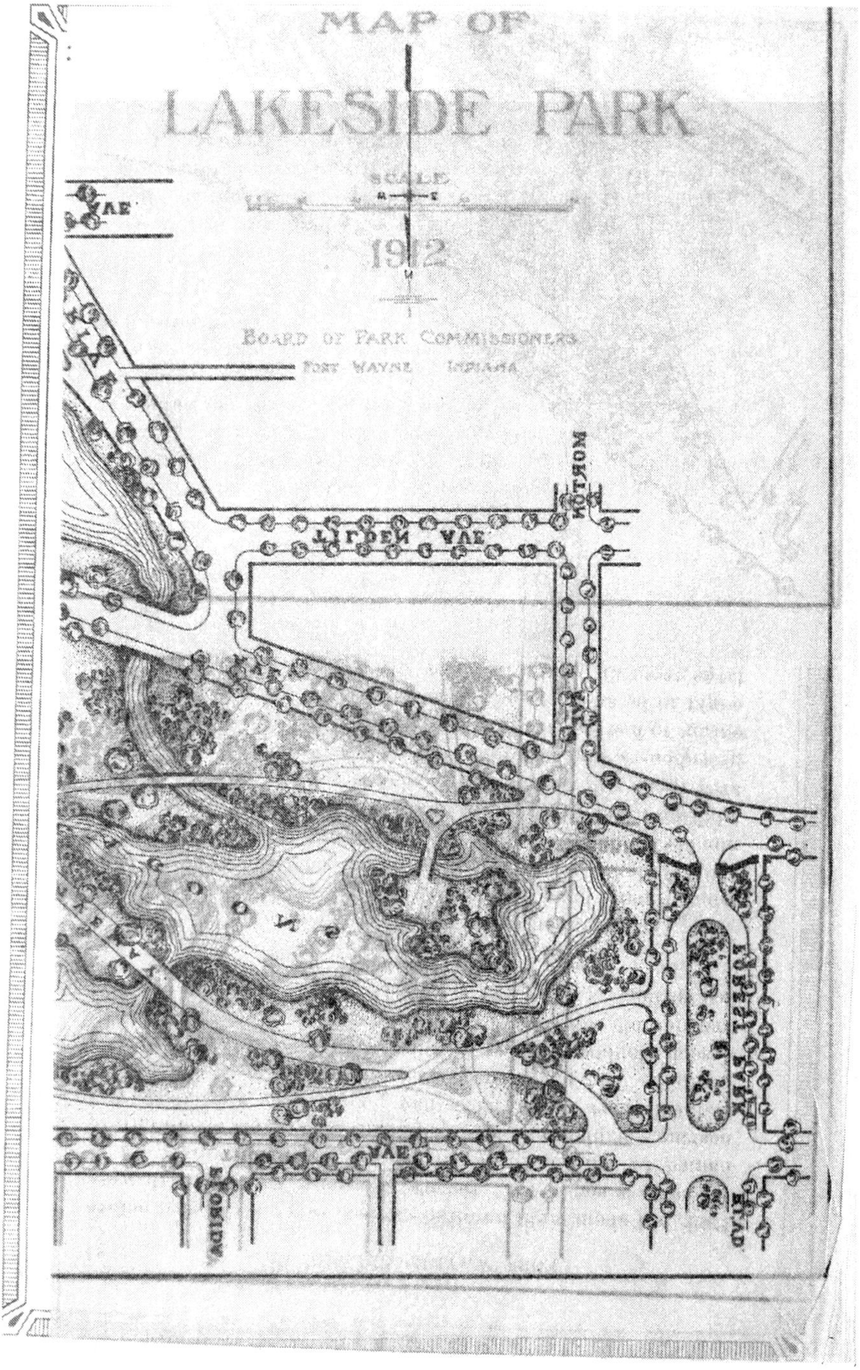
MAP OF
LAKESIDE PARK
SCALE
1912
BOARD OF PARK COMMISSIONERS
FORT WAYNE INDIANA
TILDEN AVE
MORTON

and had them placed in this park. The Wildwood Builders Company constructed a speaker's stand for the dedication exercises on July 4th, and afterwards donated the same to the Park Board, to be utilized as a band stand. Several benches and tables for the use of picnicers were also constructed in this park by this department.

HAYDEN PARK.

The old gravel walks in this park were torn out and replaced with new walks prepared with stone screenings mixed with liquid asphalt and rolled. The park is in a densely populated section, and in close proximity to a school, hence it is much traversed. Many people have acquired the unfortunate habit of walking on the grass along the edges of the walks. Several hundred shrubs will be planted in this park next spring.

WEISSER PARK.

Early last spring we removed a large tree which stood within the baseball diamond. A tennis court was constructed, and a number of benches and tables erected. Some earth was secured when Weisser Park Avenue was paved, to make a fill in the park along Hanna Street. Weisser Park Avenue is now paved up to the park. A comfort station ought to be constructed in this park.

RESERVOIR PARK.

A large number of shrubs were set out in this park during the past season, but many more will be required before this work is completed. Reservoir Park, during the winter months, is used more than any other park in the city, on account of the large number of people going to this park to enjoy skating, and hundreds of children use the hills for coasting.

OTHER PARKS.

The trees and shrubbery in Williams Park are growing nicely, several hundred new shrubs being planted there last fall. Hirons Park, located on Piqua Avenue, was laid out and planted according to the plans drawn by Mr. W. H. Hillary. A large amount of grading was done in Pontiac Place

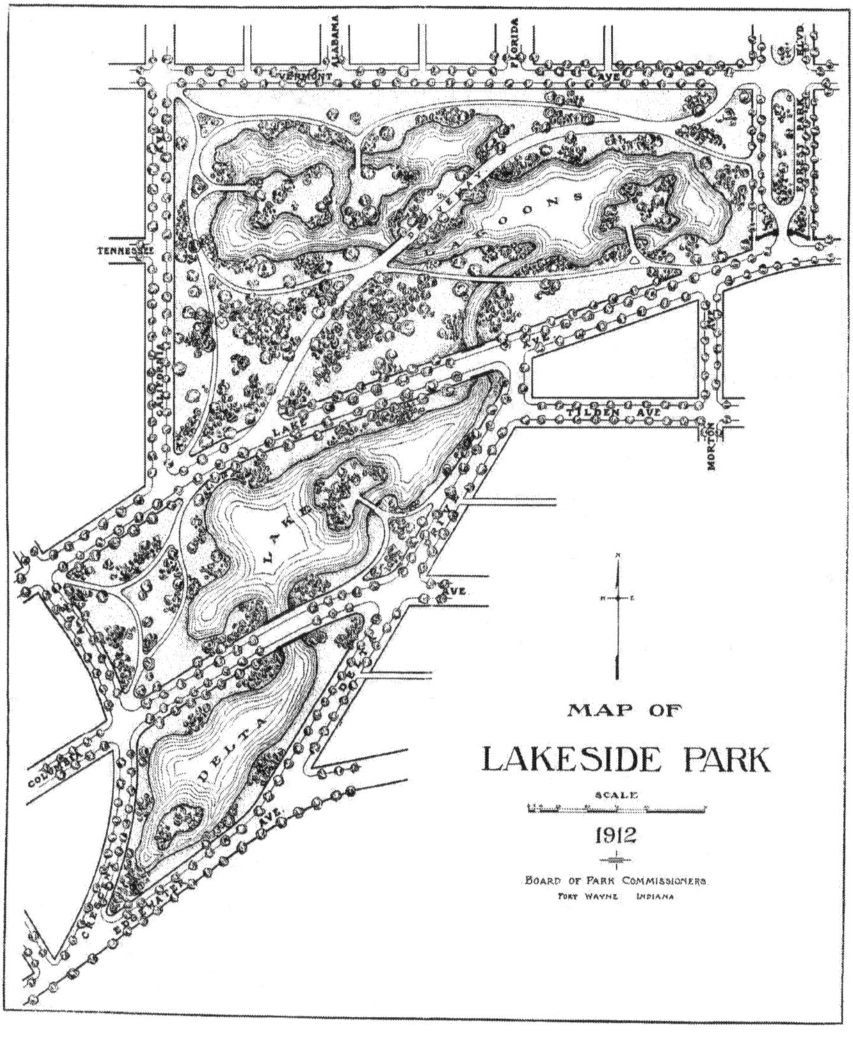

MAP OF
LAKESIDE PARK
SCALE
1912
BOARD OF PARK COMMISSIONERS
FORT WAYNE INDIANA
VERMONT
ALABAMA
FLORIDA
AVE
TENNESSEE
LAGOONS
TILDEN AVE
MORTON
LAKE
DELTA

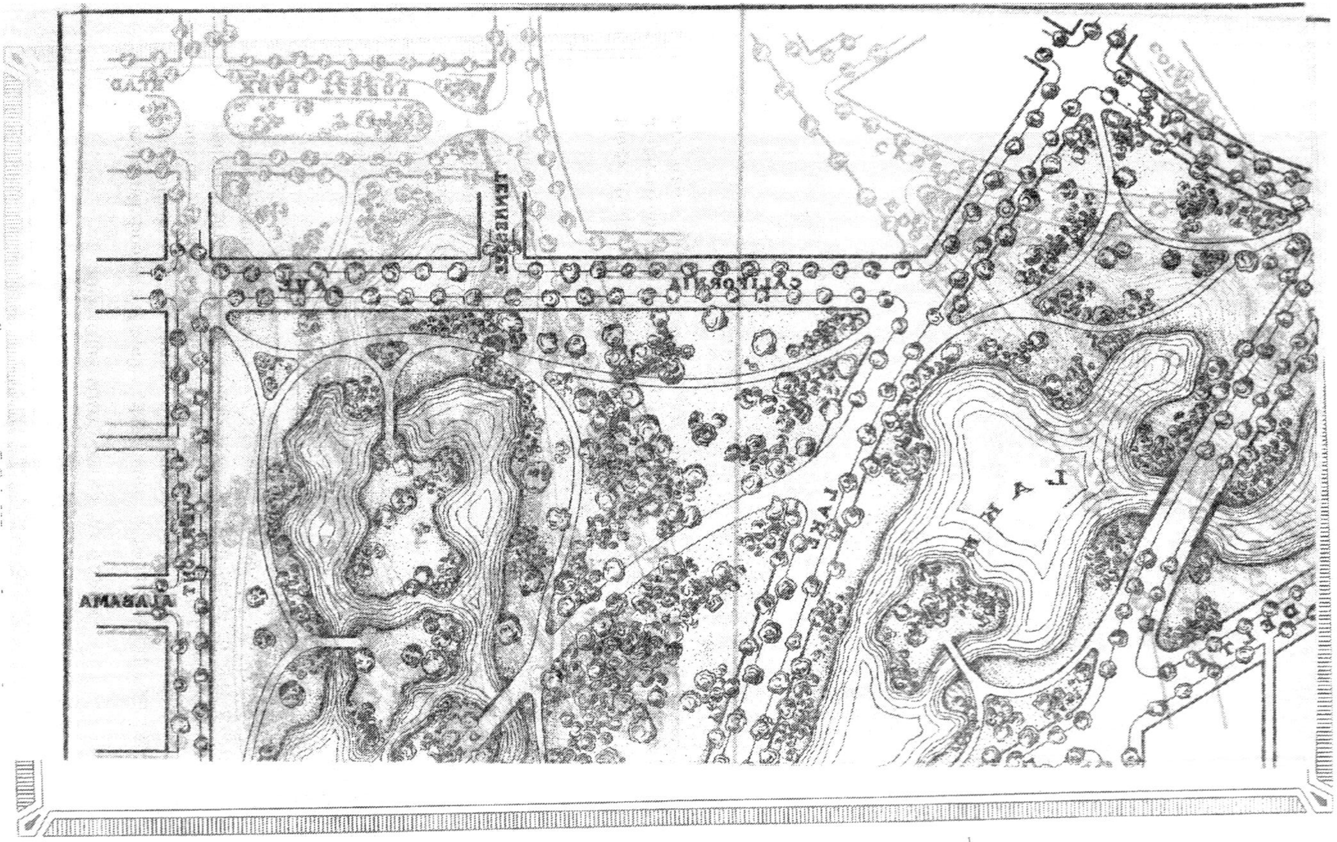
ALABAMA
CALIFORNIA
TENNESSEE
VERMONT
FOREST PARK BLVD.
LAKE

Park, but about eight hundred yards of earth are needed before the same is completed. Several thousand loads of earth were hauled to Thieme Drive during the year, and this work is nearing completion. The usual care was given to McCulloch and Old Fort Parks. The flag pole in Old Fort Park was painted. The parkstrips along Edgewater Avenue and St. Joseph Boulevard were kept in their usual good order, and considerable work was done on the Guldlin playgrounds early last spring.

GREENHOUSES.

Our greenhouses are now in good condition after several minor repairs have been made. During the past season about 150,000 plants were raised, and in order to care for this large number of plants, we used the twenty-four large hot-beds in addition to the greenhouses. On account of the increased park area there will be a greater demand for the product of our greenhouses, and it would be a wise investment for a greater output to increase the efficiency of our plant. One large section ought to be added to our houses for the better housing of our large tropical plants, in order that the space of the present houses might be utilized for the propagation of the smaller plants. In regard to the heating plant, the department very earnestly recommends the installation of a new battery of boilers.

NURSERY.

The nursery at Lawton Park, which has been one of the smaller parts of the city parking scheme, was considerably enlarged last spring. Several thousand tree and shrub seedlings were purchased and set out. The department will be enabled to raise a sufficient number of trees and shrubs needed for its own use. During the past season 4,500 shrubs raised in the nursery, were transplanted in the parks. This constitutes a great saving, as trees several years old cost from a dollar up, the seedlings being purchased at a very low figure. Our nursery will prove to be an excellent investment.

ROAD OILING.

The roadways in all the parks were oiled with number four road oil, and were in fine condition during the entire season,

A Corner of the Lake in Swinney Park.

thereby greatly abating the dust nuisance, which is very detrimental to plant life. Several streets were oiled for private parties during the past summer, however, it is the opinion of this department that this work should be done under the supervision of the Street Commissioner. The streets ought to be graded and leveled before the oil is applied in order to get good results, and that work properly belongs to the Street Department.

CONCLUSION.

In concluding this report I want to express my appreciation for your hearty support and encouragement in my work, and thank you for the consideration given to the recommendations made by me. It is indeed a pleasure to carry forward the work under your intelligent and well directed efforts to beautify our park lands.

Respectfully submitted,
A. W. GOERS,
Superintendent.

Tree Injured by Grading for Sidewalk. Saved, Trimmed and Treated by Our City Forester.

Fort Wayne, Indiana, December 31, 1912.

Board of Park Commissioners, Fort Wayne, Indiana:

GENTLEMEN:—Herewith I submit the first annual report of the Forestry Department for the year 1912. This being the first report of the activities of this new department, it would seem appropriate to briefly review its inception and development.

On March 12, 1912, the Board of Park Commissioners' Shade Tree Ordinance was unanimously passed by the Common Council and duly signed by the Mayor.

The Department was established April 2nd, 1912.

EARLY ACTIVITIES.

As soon as the Fort Wayne newspapers announced that the Board of Park Commissioners had appointed a City Forester, much interest was manifested on all sides. The Secretary was at once deluged with inquiries and requests for the Forester's services, and within two weeks after his arrival, actual field work was begun. Fort Wayne young men were trained in the various processes of practical shade tree preservation, and this expert service was placed at the disposal of the public. This work consisted of eradicating and controlling tree diseases by the employment of power sprays; the symmetrical trimming of street, shade and lawn trees, the pruning of fruit trees; planting and transplanting of shade trees; tree surgery, etc. The men employed on the force were taught the details of each operation and have now become very efficient.

THE FORESTRY FORCE AND EQUIPMENT.

At present we have two forces of trained men, each in

charge of a competent foreman. The equipment for each force consists of a large single horse wagon, and complete paraphernalia for doing every branch of the work. The tools are of the best and most approved type, thus enabling the work to be executed quickly and efficiently; the charges for the work thereby being placed at a minimum.

INSTRUCTION PAMPHLET AND PUBLICITY.

Acting upon the order of the Board of Park Commissioners to prepare a general instruction pamphlet, a booklet of twelve pages was issued. This serves as general information to the public, and is distributed freely. It contains the State Shade Tree Law; the City Shade Tree Ordinance; rules and instructions governing planting, trimming and removal of trees along sidewalks; specifications for removing trees and cutting tree roots by sidewalk contractors; practical spraying, pruning and tree surgery instructions; the mode of securing a pruner's license, and the manner in which the services of the Forestry Department may be secured. This places accurate information in the hands of interested citizens at but a slight expense, and the result of this education was manifested in many ways.

PUBLIC FORESTRY WORK.

Tree Improvement Resolution Number One, after taking its due, legal course, was adopted September 21st, 1912. It embraced, "The removal of all dead and dying trees, and trees improperly placed; to prune and preserve all trees requiring the same, and doing all other things necessary to put the trees in first-class condition on East Wayne Street, from the East line of Barr Street to the West line of Grant Avenue, and Washington Boulevard East, from the East line of Barr Street to the West line of Glasgow Avenue; also to plant good, healthy nursery grown Norway Maples uniformly in all open places along both sides of East Wayne Street, from the East line of Lafayette Street to the West line of Harmer Street, and along both sides of Washington Boulevard East, from the East line of Barr Street to the West line of Harmer Street, except in front of business places, along said streets, where the sidewalks extend to the curb, and to replace free of charge the trees so planted, during the first two years, if they fail to live."

This improvement is now in progress, and will be completed within a short time. This work is a revelation, and has added 20 per cent to the appearance of the two streets. Very favorable comment has been passed upon this work, and it demonstrates that systematic work of this character is a step in the right direction and meets with the hearty approval of property owners.

Many trees along the river bank on St. Joe Boulevard, which were dead or had dangerous dead limbs upon them were removed under the direction of the Park Board, and the live trees properly trimmed, greatly adding to the appearance of that mile of river bank.

STREET TREE PLANTING.

On October 5th, 1912, the Forester was directed to plant Oriental Plane trees along the East side of Anthony Boulevard, from Pontiac Street to Wayne Trace. This required 151 trees, guards and stakes. The trees were planted in a staggered arrangement, according to the standard boulevard scheme. This arrangement gives the boulevard a very dignified appearance, as well as making a delightful shady sidewalk for pleasure seekers when the trees are grown. This is the first boulevard planting in Fort Wayne, and from present indications these boulevards will develop into drives and walks of great beauty and utility. The Oriental Plane tree has been exclusively reserved for boulevard planting, and the beauty and dignity of this tree, as well as its hardiness and longevity commends it for this purpose.

On November 16th, 1912, the Board granted the petition of the property owners to plant Norway Maples on Oliver Street, from Pontiac to McKee Streets. This work was started and will be completed this spring. The trees are being planted thirty feet apart, and properly protected with tree guards and stakes. This is a very commendable work and will develop into a uniform row of beautiful shade trees.

PRIVATE FORESTRY WORK.

During the nine months of its existence the department has executed 217 private contracts. The cost of this service varied from 45 cents to $69.79. This work consisted of spraying for tree diseases; planting; trimming shade trees; pruning fruit trees, hedges and vines; tree surgery; tiling and fertilizing trees, etc. That this feature of the department's work was popular is evidenced by letters of approval and the re-orders for 1913. The scale price of this class of work has been set at 35 cents per hour per man, and 17 cents per hour for horse and wagon; to these figures are added the cost of materials and supplies used. This figure is as low as is consistent with the high standard of work and service rendered, and, based on the amount of work done in 1912, is practically the net cost to the department for doing this work.

LICENSED PRUNERS.

Two residents of the city have satisfactorily demonstrated their competency, and were licensed to engage in the business of trimming, pruning, planting and treating trees. This feature protects the citizens from the tree butchers who have been mutilating our beautiful city shade trees in years past. This class of work can be done most satisfactorily by young men with ability to climb, and old men who applied were induced to secure employment for which they were better adapted. At present there is a very healthy co-operation between the Forestry Department and the licensed tree pruners and tree agents; and working thus in harmony, the best results will be attained.

INSECTS AND TREE DISEASES.

Fort Wayne is very fortunate in having but few tree diseases. The scale insects seem to be present in the greatest numbers; the San Jose Scale heading the list, with the Oyster Shell Scale ranking second. The Carolina cottonwood, or common poplar trees, serve as breeding places for this latter scale, and a removal of all the poplar trees in Fort Wayne would almost solve the Oyster Shell Scale and the Scurfy Scale problems. The Cottony Maple Scale and the Wooly Maple Scale are present in small quantities and do but little harm. An annual spraying of all the fruit trees upon city lots would result in greatly diminishing the scale, and prevent its spread to the shade trees. The scale insect has also been noticed on the shrubs in the parks and in some instances in

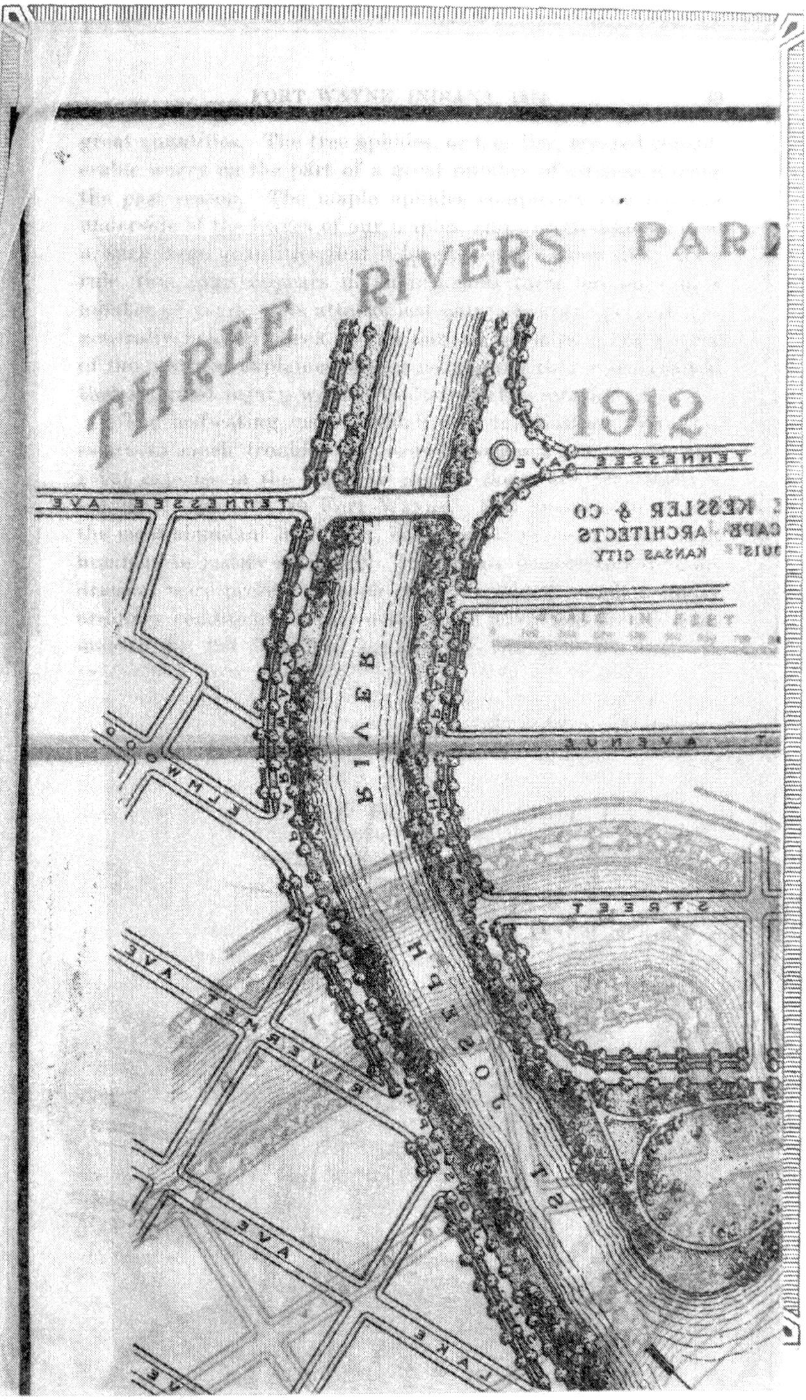
THREE RIVERS
1912
TENNESSEE AVE
KESSLER & CO
ARCHITECTS
KANSAS CITY
ST JOSEPH RIVER
STREET
LAKE

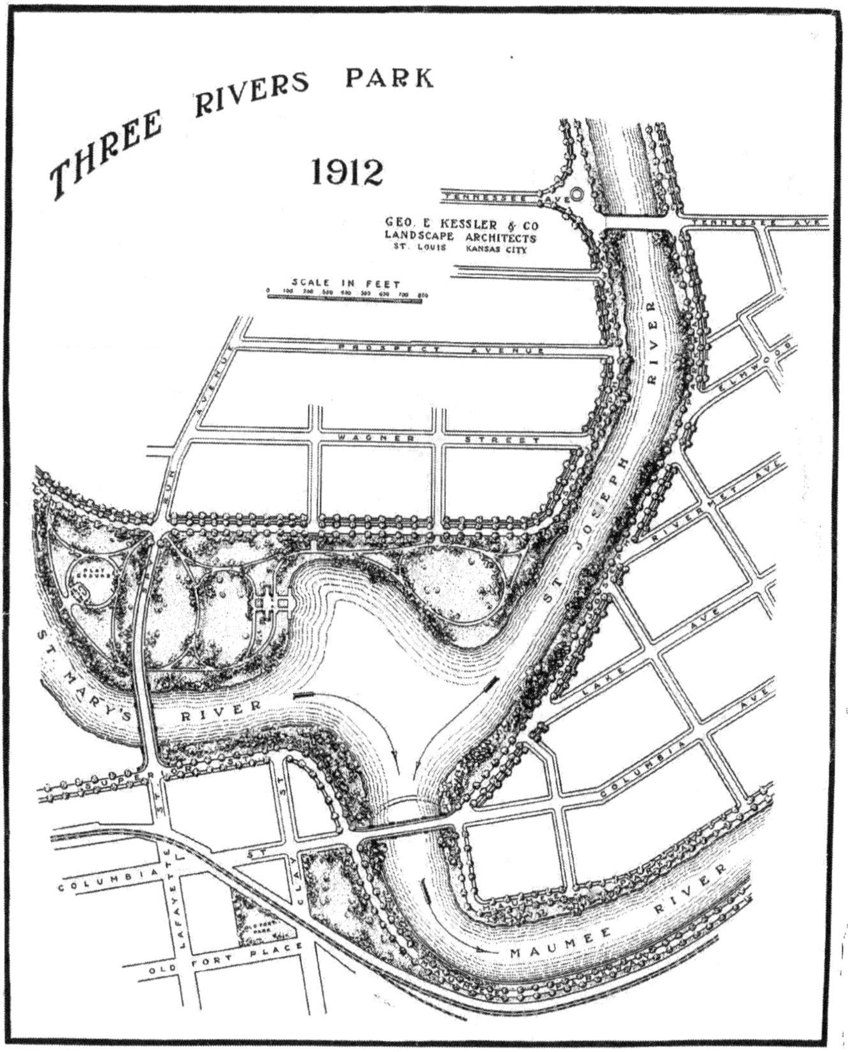

THREE RIVERS PARK
1912
GEO. E KESSLER & CO
LANDSCAPE ARCHITECTS
ST. LOUIS KANSAS CITY
SCALE IN FEET
TENNESSEE AVE
TENNESSEE AVE
PROSPECT AVENUE
WAGNER STREET
ELMWOOD
LAKE AVE
COLUMBIA AVE
ST JOSEPH RIVER
ST. MARY'S RIVER
MAUMEE RIVER
SUPERIOR
LAFAYETTE ST
CLAY ST
COLUMBIA
OLD FORT PLACE

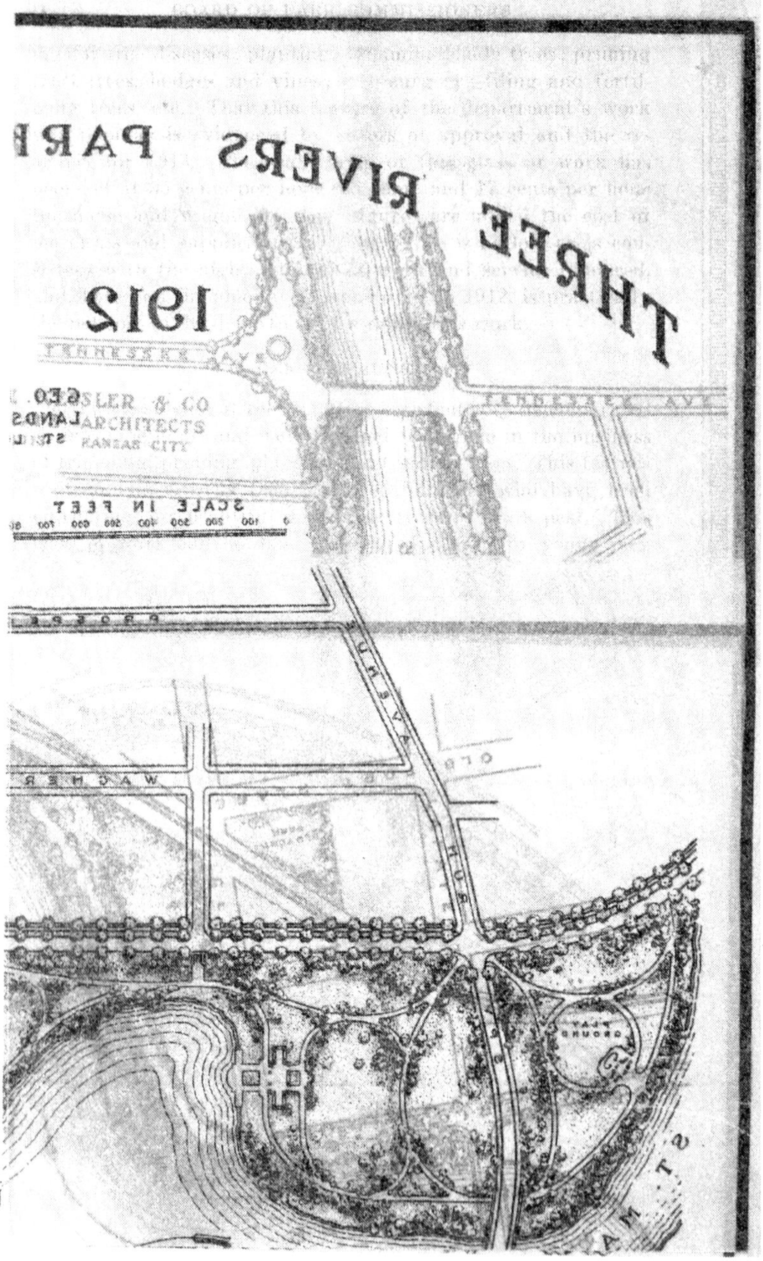
THREE RIVERS PARK
1912
SCALE IN FEET
KANSAS CITY

great quantities. The tree aphides, or tree lice, created considerable worry on the part of a great number of citizens during the past season. The maple aphides completely covered the underside of the leaves of our maples, and excreted honey dew in such large quantities that it blackened the sidewalks. As a rule, this aphis appears in an injurious form but once in a number of years. Its attacks last only a few weeks, and it is generally held in check by its natural enemies. The nature of the pest was explained to our people and they were assured that no great injury would result from the attack.

The leaf-eating insects which give the Eastern City Foresters so much trouble, and cause enormous tree losses and great expense in the effort to control them, are practically a negligable quantity in Fort Wayne. The Tussock Moth was the most abundant leaf eater, but was not present in sufficient numbers to justify spraying. The boring insects and fungous diseases were present in such amounts, as are common under ordinary conditions. Their destructive work was greatly augmented by the improper pruning of the tree butcher. A triplex power pump is used by the department to spray trees badly infested with scale. In certain localities the shade trees are slowly dying from the effects of the scale, but a timely spraying of these trees will hold the scale in check and save the trees. This feature of the work of the department will be developed to a somewhat greater extent during the coming year.

ARBOR-DAY.

On October 25th the Forester assisted two public schools in their Arbor-Day celebration. During the afternoon of that day the forestry force planted nineteen American elm trees in the James H. Smart School playground. This work was a donation to the school by Messrs. Griffith and Fair, architects of the building. The children assembled in the playground at two o'clock P. M., and after singing a few appropriate arbor-day songs, witnessed the planting of a tree, the Forester explaining the details of each step, as well as the natural function of each component part of the trees. At three o'clock P. M. of the same day the Forester addressed the children of the Miner School. In the future a more earnest co-operation

Our City Forestry Department Transplanting Large Tree.

will be attempted between the public schools and this department, in an effort to create a keener interest in trees, shrubs and plant life.

THE MUNICIPAL NURSERY.

Early last fall a municipal nursery was started, where seedling trees will be grown from seed, and other propagation work carried on. This nursery will care for about four thousand trees, and will shortly turn out six hundred trees annually, of the proper size for planting along our streets or within our parks. The trees, when ready to be set out, will have a value of $1.00 each. All approved, standard varieties of shade trees will be grown. The nursery now contains some two hundred and fifty trees, about one-half of which will be large enough to plant in the spring. The production rotation of six hundred trees annually will be started the coming year. The ground utilized for the nursery is located South of Spy Run Creek, in Irvington Park Addition, purchased by Park Commissioner Louis Fox, and held in trust by him for the city, to eventually become a part of John H. Vesey Park, which lies immediately North of this tract of land. The action of Mr. Fox, in giving consent to use this ground for the nursery, should be commended.

ILLUSTRATIONS.

Your attention is called to the views, illustrating some of the work done by this department. A large Basswood tree in front of the property owned by Mrs. J. W. Foley, opposite the Market House, the trunk of which was badly decayed, was treated by removing the diseased portions, after which the cavity was thoroughly disinfected and filled with about one ton of re-inforced cement mixture. This picture shows the trunk of the tree before, during and after treatment. This class of work is similar to dentistry, and the effectiveness depends upon its thorough execution.

A number of large trees were transplanted during the fall. This class of work can be done very successfully if properly executed. Special attention is called to that part of the illustration showing the many roots left intact in transplanting a large tree. This particular tree, shown in the picture, was

A Diseased Tree, Showing Defective Wood.

All Decayed Parts Removed, Ready for Filling.

New Bark Gradually Heals Over the Cement.

planted in front of the residence of Mr. C. F. Bicknell, on Wildwood Avenue, and is a Norway Maple about fourteen years old.

A large Maple tree in front of the residence of Mr. Adolph Katzenberg on Columbia Avenue, the roots of which were exposed by a change in the grade when a new sidewalk was constructed, thereby greatly lessening the vitality of the tree, was treated by pruning and constructing a cement coping around it. The roots were thoroughly fertilized and covered with rich soil to its original depth. The picture shows how a tree responds to intelligent treatment, and also demonstrates the manner in which a tree develops, if properly spaced, the spread of this tree being forty-five feet.

DEPARTMENT SELF-SUSTAINING.

An endeavor was made to make the department self-sustaining. This has been accomplished, as will be noted in the summary herewith submitted:

Cost of operating the department		$ 3,407.19
Private contracts	$2,339.91	
Public work	129.27	
Preparing nursery in Vesey Park	211.33	
Nursery stock in nurseries	72.15	
Tools and equipment	223.55	
Anthony Boulevard planting	380.67	
Oliver street planting	75.90	
Wayne and Washington, general charge	135.35	
Balance		160.94
	$3,568.13	$3,568.13

CONCLUSION.

In closing my report I desire to thank the Board of Park Commissioners for the encouragement and support given; the city officials, office force, the newspapers and the public in general, for the many kind favors rendered, and the spirit of co-operation that has been shown this department in carrying forward its work. A continuance of these relations is earnestly desired and anticipated.

Respectfully submitted,

CARL J. GETZ,

Forester.

Manual Training and Public High School, Fort Wayne, Indiana.

PARK HISTORY

A BRIEF history is here given of the park system in our city, and its evolution. Fort Wayne has acquired much of her park area by gifts from her public spirited citizens, and our people have been very grateful for this generous spirit. They have expressed their appreciation by giving such donated park areas the names of its donors. Regarding the park lands that have been purchased, time has shown that no better investment could have been made, as these properties have not only enhanced the beauty of our city, but have greatly increased in value and have become a large part of its assets.

FIRST PARK IN 1863.

The development of our park system covers a period of fifty years. The beginning was made, when in 1863 the city purchased as its first park site fractional Lot Number Forty Cyrus Taber's Addition from Mr. Harry Seymour for $800.00. This small parcel of ground was given the name of Old Fort Park, as it was a part of the site upon which General Anthony Wayne erected his fort in 1794. The ornamental iron fence surrounding the park, and the handsome flag pole was the gift of Mr. Henry M. Williams.

THE SECOND PURCHASE.

The lands comprising the major portion of Lawton Park, the original name of which was City Park, which later was changed to Northside Park, and still later to its present name, in honor of General Henry W. Lawton, were purchased on January 24th, 1866, from Wm. Fleming, S. B. Bond, C. D. Bond, W. H. Jones, Hugh B. Reed, Henry J. Rudisill and J. W.

Old Fort Park, a Section of the Original Fort Grounds.

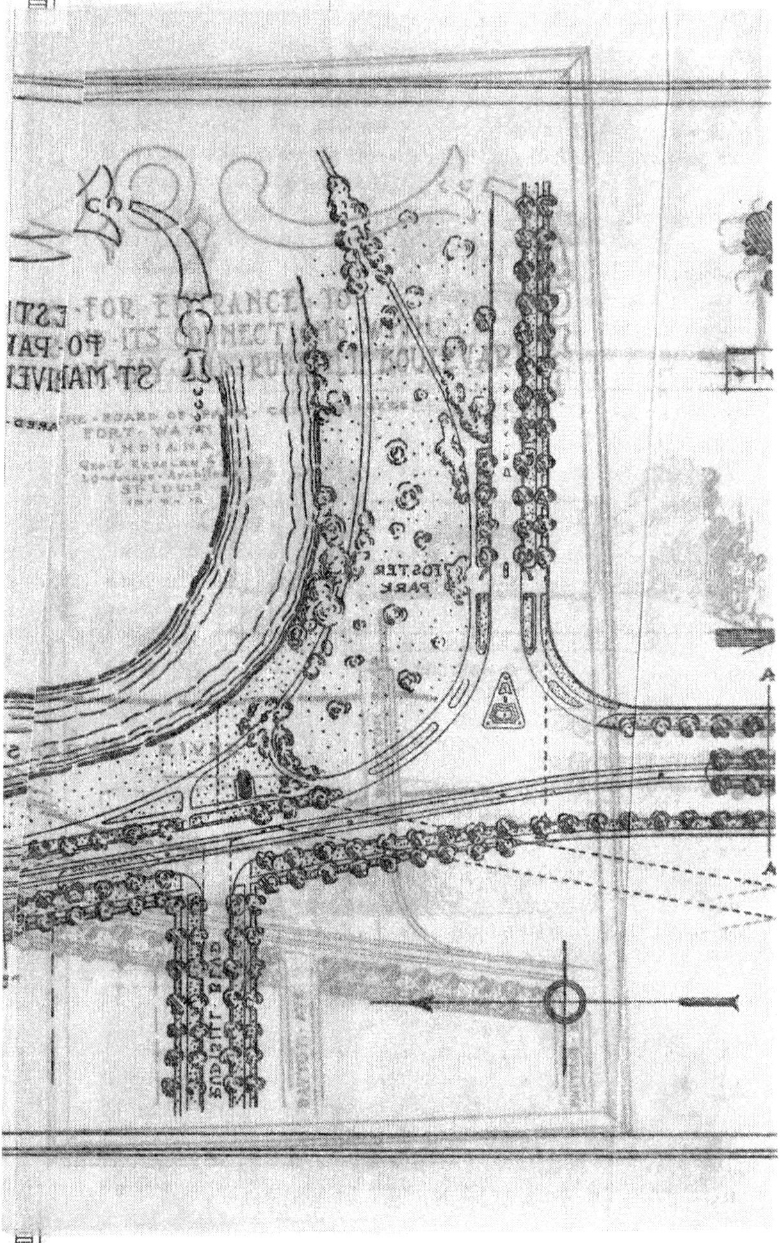
FOSTER PARK
INDIANA

Old Fort Park, a Section of the Original Fort Grounds.

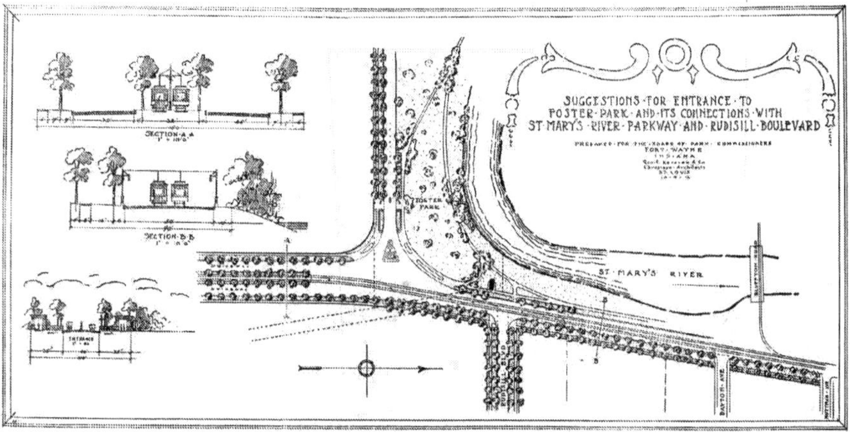
SUGGESTIONS · FOR · ENTRANCE · TO
POSTER · PARK · AND · ITS CONNECTIONS · WITH
ST · MARY'S · RIVER · PARKWAY · AND · RUDISILL · BOULEVARD
FORT · WAYNE
INDIANA
ST. LOUIS
SECTION · A · A
SECTION · B · B
POSTER PARK
ST · MARY'S RIVER
RUDISILL BLVD
BAYTON · AVE.

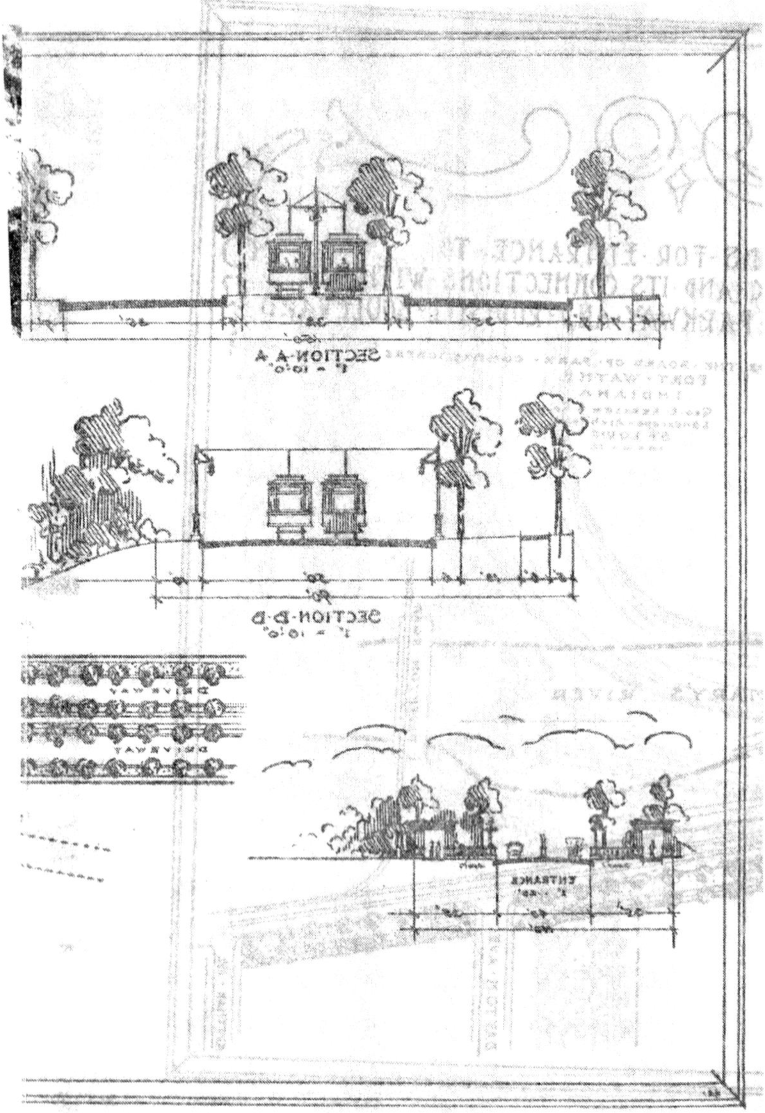

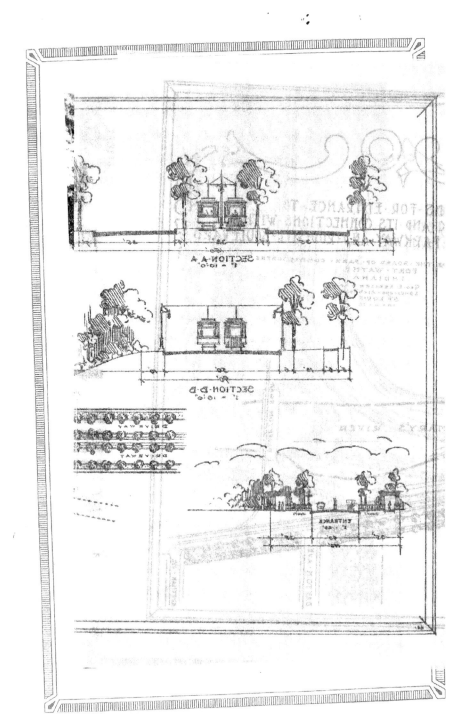

Dawson, at a cost of $35,500.00, for which amount the city issued bonds. The balance of the ground was purchased in 1866 and 1881 from Mathias Mettler for $1,755.00, making the original cost of the park $37,255.00.

In 1872 the City Council donated to the Fort Wayne, Jackson & Saginaw Railroad Company twenty acres of the West portion of the park for a right of way and shop purposes. The balance of the ground was platted into city lots, and the park abandoned. Fourteen full lots and two half lots were sold for $8,600. Later, as there was no demand for the lots, the plat was vacated and the land again developed into a park, the North portion of it being used for the City Water Works pumping station.

In 1881 the city began to re-purchase the lots that were sold, and since that time six full lots and two half lots have been secured at a cost of $6,176.00, the two half lots being used for the opening of Clinton Street. The purchase price of Lot Eighty-three was $1,800.00, including the house upon it, which house was later sold for $315.00, and removed. The total cost of the land now included in the park has been $34,516.00, with seven full lots and two half lots remaining to be purchased. The present area of the park is 31.50 acres.

THE FIRST DONATION.

The city secured the site of the present Hayden Park in September, 1866, by the purchase of six lots in Hanna's Partition Addition, Plat B, from H. W. Hanna, S. D. Hanna, Jessie E. Bond and C. H. Hanna, at a cost of $4,500.00.

Eliza Hanna Hayden donated Lot Number Six in the same addition, on condition that the park be named and known as Hayden Park. The area of the park is 1.11 acres.

SITE FOR RESERVOIR.

Reservoir Park was acquired through the necessity of obtaining a suitable site for the location of a reservoir for our water works system. Sixty lots in Hamilton's Fourth Addition were purchased in 1880 from Andrew and Montgomery Hamilton, Mary H. Williams and Ellen and Margaret V. Hamilton, for $24,000.00. The park contains 13.12 acres, a small portion

Public Library, Fort Wayne, Indiana.

of which is used by the City Water Works for its reservoir embankment.

MORE EARLY DONATIONS.

McCulloch Park was the first park the total area of which was the gift of one of our public spirited citizens. Ex-Secretary of the Treasury Hugh McCulloch and his wife Susan, donated this park in 1886, on condition that it be named and known as "The McCulloch Park." The park has an area of 4.07 acres.

In 1890 Henry M. Williams and Mary Hamilton Williams, his wife, donated Lot Number Fifty-nine, South Calhoun Street Addition, for park purposes. The Board of Park Commissioners by resolution, named it Williams Park in honor of the donors. The park is located at the junction of South Calhoun Street and Piqua Avenue.

ORFF PARK.

This miniature park was secured in 1892, when the city purchased from John Orff a tract of land lying between the North line of Old Main Street and the North line of the present Main Street, from Rockhill Street East to a point where the two Main Streets intersect, at a cost of $5,500.00. Most of this ground was used for the relocation of Main Street, leaving a pretty spot as an approach to the Main Street bridge.

SWINNEY PARK.

In his last will, Thomas W. Swinney, deceased, devised to the City of Fort Wayne, in fee simple in trust for a public park, under certain conditions, the property now comprising the major portion of Swinney Park. The will contained a provision that his children should enjoy the use of the grounds so long as either of them should live.

On June 1st, 1893, a contract was entered into between his children and the City, whereby the City secured immediate possession of the property by paying a rental of $600.00 per year and taxes, and under it the City proceeded with its park improvements. The contract was made for a period of thirty years, terminating in 1923, or sooner, if the life tenancy of his children should cease. At the expiration of the contract the

Guldlin Playgrounds, Boys' Section.

city has the right to extend the contract for an additional term of twenty years.

In 1869 the City purchased and added to Swinney Park eleven and one-half lots in Ninde's First Addition, from Samuel Cary Evans, for $7,000.00, and one and one-half lots in the same Addition, from Humphrey W. Y. Porter, for $1,200.00. The entire area of Swinney Park is 45.30 acres, and the total cost, including the rental, to date is $20,200.00.

GULDLIN PLAYGROUNDS.

The Guldlin playgrounds, although not then known by that name, were purchased as a site for number two Water Works pumping station. The City purchased better than it knew when this land was secured in 1897. A temporary pumping station was erected, but was abandoned and removed the following year when additional land was purchased on the opposite side of the river, upon which a large pumping station was erected.

The cost of the entire tract was $7,011.67, and was purchased from the Rockhill Heirs and from N. D. Doughman, Charles McCulloch, Charles Pape, Charlotte Wefel and Frederick and Henry Bade. The playgrounds were graded and equipped in 1911, at a cost of $7,000.00, of which sum the City gave $2,500. The balance was raised by private contribution, of which a very large part was given by Mr. and Mrs. O. N. Guldlin, after whom the grounds were named.

LAKESIDE PARK.

The original part of Lakeside Park was acquired in 1908. It was purchased from the Fort Wayne Land and Improvement Company, at a cost of $5,000.00, while a considerable portion was donated, The Fort Wayne Forest Park Company in 1908 giving Blocks Three, Four and Five, Forest Park Addition, and the Boulevard Realty Company donating Lot A, Forest Park Place, and also the parkstrip in the center of Forest Park Boulevard. The latter donation was made in 1912. The City, in 1912, purchased the Parham property for $2,800.00, making the total cost of the park $7,800.00. In 1912 this park was improved at a cost of $17,500.00, which amount was raised by

View of the St. Joseph's River.

special assessments. The area in this park, including the parkstrip, is 26.93 acres.

THIEME DRIVE.

The property along the East bank of the St. Mary's River, from Main Street to Washington Boulevard West, was acquired by condemnation proceedings of the Board of Public Works in 1908 and 1909, at a cost of $3,578.55, which amount was raised by special assessments. This parkway was named Thieme Drive by the Common Council at the recommendation of the Park Commissioners, in honor of Mr. Theodore F. Thieme, who very generously made a handsome improvement at the North end of this drive, at a cost of $7,000.00. The improvement of this parkway is being gradually done by the Park Board out of its regular revenues.

WEISSER PARK.

Weisser Park cost the city $10,500.00. It was purchased in 1909 from Magdalena Weisser for the above consideration. The park contains 15 acres, and is the most beautifully wooded piece of park property that the city owns. It needs very little artificial transformation. The Park Commissioners intend to eventually add more land to this park, which will double its size.

PARKSTRIPS.

During the year 1909 the City came into possession of all the unplatted land lying between Edgewater Avenue and the Maumee River, containing 15.42 acres, by donation of the Fort Wayne Land and Improvement Company, and in 1911 the same Company deeded to the City the land between the St. Joseph River and St. Joseph Boulevard, from Columbia Avenue to Tennessee Boulevard, the City paying a paving charge of $511.75, which was a lien on the property. There are 3.07 acres in this area. During the same year Henry R. Freeman donated 1.20 acres of land lying between the St. Joseph River and Elizabeth Street, from the alley East of Lots 1-2-3, Freeman's Amended Addition, to the alley South of said Addition.

Early in 1912 the City purchased from Howell C. Rockhill 2.95 acres in Butcher's addition, lying East of Mechanic

Japanese Pavilion at Swinney Park. Fort Wayne, Indiana.

Street, West of the St. Mary's River, and South of Fair Street, the consideration being $1,284.83. Later during the same year two parcels of land, containing one acre, lying between Broadway and the St. Mary's River, South of the Bluffton Road, were donated by Montgomery Beaver et al., and R. L. Romy and S. S. Fisher.

RECENT DONATIONS.

During the year 1912 Fort Wayne came into possession of four additional parks, the most important of them being Foster Park, lying along the banks of the St. Mary's, donated by Col. D. N. Foster and Hon. S. M. Foster. The park contains 63.91 acres, considerably the largest park in the city. The donors are purchasing forty-six additional acres, which will extend Foster Park to the Stellhorn bridge, which ground they also will donate to the city for park purposes.

The late John H. Vesey had planned to donate to the city a plat of ground containing five acres in his addition, known as Irvington Park. After his death his widow deeded the property to the city, on condition that the park be given the name of "The John H. Vesey Park."

Pontiac Place Park, so named by the Board of Park Commissioners, was donated by the Pontiac Place Company, who laid out an addition adjoining this property. It contains 2.06 acres, and is located on Pontiac Street, East of Anthony Boulevard.

Albert R. Hirons donated Lots 83-84-85 in Hirons' Addition, to be used for park purposes. It is located on Piqua Avenue, corner of Fairfax Street. The park was named in honor of its donor.

ROCKHILL PARK.

In 1908, when the Commercial Addition in Westfield was platted, the owners set aside 7.57 acres for park purposes. The conditions under the dedication are that the city will come into possession of this park whenever Westfield is annexed to the city.

HOLDING IN TRUST.

In order to prevent two excellent pieces of park properties being lost to the city, as the Park Fund did not admit of their

Entrance to Lakeside Park, Fort Wayne, Indiana.

purchase at the time, Park Commissioners Foster and Fox have purchased and are holding these properties in trust, to be purchased by the city as soon as funds are available; Mr. Foster having purchased from the Hamilton National Bank 9.65 acres for $8,000.00, lying between the St. Mary's River and the Lake Shore and Michigan Southern Railroad, West of Wells Street, and Mr. Fox purchasing from the John H. Vesey Estate 9 acres adjacent to Vesey Park, for $5,400.00. The latter property when purchased by the city will be added to Vesey Park.

GRAVEL PITS.

In 1864 the city purchased from Julian Benoit Lots Number 11 and 12, Jones' Addition, for $850.00, and in 1887 Lot Number 13, in the same Addition was purchased from Ed. P. Williams, et al., for $1,000.00. This property was used as a gravel pit, but later was filled, graded and platted into four lots, known as City Subdivision. The Park Board recently erected a large barn for the Park and Forestry Departments across the rear of these lots, the front of which will be made into a miniature park.

The gravel pit lying between Zollinger Street and Dwenger Avenue, in White's Fourth Addition, was purchased on June 16th, 1873, from James B. White, for $15,000.00, the city issuing its bonds for the full amount. The tract originally contained five acres, and was purchased for a fairgrounds. In the opening of Dwenger Avenue a small portion of this tract was used, the area of it now being 4.95 acres. This ground is a nucleus to a future park in the Eastern section of the city.

PARK AREA.

Fort Wayne now has 227 acres of park and parkstrip area. This is a record to be proud of, especially since 72 acres of this amount have been added during the past year. There is probably not a citizen of Fort Wayne who would trade his city with its beautiful parks for a city which lacks such necessary conveniences. It is not merely a matter of appearance, it is a matter of intelligence, civilization and health.

Compiled by
CHARLES J. STEISS, Secretary
BOARD OF PARK COMMISSIONERS.

Winter Sport at Reservoir Park.

Fort Wayne, Indiana, December 31, 1912.

To the Honorable Board of Park Commissioners:

ENTLEMEN:—The year just closed has been a period of satisfactory development in the Park Department. A large amount of improvement work has been accomplished, and extensive studies with maps and reports have been made for the extension of the park and boulevard system throughout the entire city. Demands are coming from all sections of our city for the rapid carrying out of the Kessler plan, and it is a source of great encouragement to know and feel that our citizens are becoming better acquainted with the relation of civic betterment to the city as a whole.

PARKS NECESSARY.

The vital necessity of parks is not recognized by many of those who are most in need of them. In this nerve-racking age, even a short hour of the quiet, peaceful influence of a beautiful landscape, acts as a restorative which goes very far to prevent that frayed, frazzled, irritable and finally enervated condition from which so many American men and women suffer.

Provision for recreation is as fundamental in the modern city as provision for any other basic need. Commercial and other organizations interested in the development of their cities and towns realize that the existence of a recreation plan is one of the best advertisements of a city, and adds value to real estate.

DELAY COSTLY.

A beautiful city, in which the health and comfort of its citizens are looked after, attracts desirable people as residents and taxpayers. Beautiful parks and boulevards enhance the value of taxable property, and thus yield a larger financial return to the city.

If we do not do it soon, the coming generations will have to pay dearly to restore, by artificial means, natural features which our foresight should have preserved.

A far-sighted policy is rapidly gaining ground among city officials and patriotic citizens generally, which insures favorable sites for parks before the growth of a city causes the land that ought to be retained for parks from becoming occupied by residences and factories.

To plan parks early, results in a better city, secures the land at a much lower figure, and induces the more rapid growth of the population and of the city's wealth. Moreover, it preserves natural scenery—one of the most valuable of a city's assets. True landscape architecture shuns too much artificiality, but exalts and refines nature.

We have permitted some of our most charming pieces of scenery and landscapes to be obliterated, but there still are many places that may be conserved which as yet have not been ruined by the hand of commercialism.

PARK FUNDS.

The activities of the Park Department are somewhat hampered by the small amount of funds available for park purposes. There was a time when the appropriations were sufficient because of the very small park area to be maintained, and upon which no extensive improvements were contemplated, but conditions have changed. Our park areas have been considerably increased, and they must be improved. This means that the maintenance and improvement expenditures will necessarily be increased, leaving little or no balance for the purchase of additional park lands. Therefore, if our city intends to keep abreast with the times, and secure needed additional park areas, the funds for the same must be derived from special appropriations, special assessments or by issuance of short time bonds.

RETURNS ON INVESTMENTS.

In order to secure the best returns on park investments the park areas should be employed, during the entire twelve months, so fast as their physical development will permit. In the matter of winter sports several of our parks afford unusual opportunities for spreading the gospel of more and better parks for the people who at present are a bit skeptical about the advisability of expending any more money for breathing spots. The summer sports have already been given intelligent attention, but with an additional small outlay of money the grounds could be made meccas of amusement for both young and old.

UTILITY OF PARKS.

The visitations to our parks during the past season have been greatly increased from that of former years. Many times permits for a picnic, reunion, tennis court, baseball diamond, football field, basketball banks, etc., were issued months ahead. Of picnics, socials or reunions, there was nearly an average of one per day during the summer months; many of these permits included the baseball diamonds and tennis courts, for which no special permits were issued on those days, and of which therefore our office has no record. During the year two hundred permits were issued for baseball games, twelve hundred permits for the tennis courts, and fourteen permits for football games. In addition to these there were many small neighborhood basket picnics of which no record was kept, and many other enjoyable little gatherings, such as outdoor evening suppers, etc. There are now permits issued for dates in August, 1913.

OFFICE RECORDS.

Our records are more complete this year than before, and gradually new forms are being introduced, which will further increase the efficiency of the office. The preceding pages have been taken up with the more interesting part of the report, the subsequent pages being given over to the financial statement, showing the receipts and disbursements, and a distribution of the expenditures for the past year, all of which is

Respectfully submitted,

CHARLES J. STEISS,

Secretary.

FINANCIAL STATEMENT
1912

RECEIPTS.

General Park Fund—			
Balance January 1st, 1912	$ 3,290.67		
Balance 1911, River Bank Fund			
Special App. Ordinance No. 465	2,706.53		
General Park Levy	32,487.45		
Total		$38,484.65	
Forestry Service—			
Private Contracts	$ 1,612.07		
Nursery Stock Sold	135.45		
Guards and Stakes Sold	24.05		
Total		$ 1,771.57	
Miscellaneous—			
Privileges	$ 61.00		
Oiling Streets	208.35		
Pruner's License	2.00		
Sundries	14.50		
Total		$ 285.85	
Temporary Loan Account—			
Temporary Loans	$ 5,282.08		
Total		$ 5,282.08	
DISBURSEMENTS.			
Executive Department			$ 3,178.01
Engineering Department			5,794.09
Bureau of Assessments			1,524.05
Park Maintenance			16,402.39
Forestry Department			3,407.19
Improvements, Construction and Acquisition			10,151.34
BALANCES.			
Balance December 31st, 1912—			
Temporary Loan Account			$ 5,282.08
General Park Fund, Less Temporary Loans			85.00
		$45,824.15	$45,824.15

DISTRIBUTION OF EXPENDITURES.

Executive Department—		
Salary of Secretary	$ 1,200.00	
Postage and Express	90.40	
Stationery and Supplies	95.13	
Furniture and Fixtures	485.53	
Photographs, Slides and Frames	122.15	
Books and Periodicals	23.51	
Transferring and Recording Deeds	12.90	
Telephone Rental	33.20	
Back Taxes on Acquired Property	108.76	
Advertising and Legal Notices	174.82	
Record Books	117.15	
Printing Annual Report	572.44	
Pocket Maps of Park System	17.00	
Railroad Fare and Subsistence	95.30	
Street Car Fare	19.00	
Sundries	10.72	
		$ 3,178.01
Engineering Department—		
Salary of Engineer	$ 1,430.00	
Rodman	506.52	
Labor	13.88	
Salary, Landscape Architect	2,033.30	
Incidentals, Landscape Architect	175.01	
Inspectors	330.00	
Topographical Map, Bal. due A. W. Grosvenor	1,130.00	
Material and Equipment	145.15	
Blue Prints	14.48	
Express Charges	6.75	
Sundries	9.00	
		$ 5,794.09
Bureau of Assessments—		
Salary of Chief	$ 600.00	
Clerical Force	328.00	
Stationery and Supplies	96.40	
City Maps	10.75	
Elliott-Fisher Writing Machine	375.00	
Assessment Records	112.50	
Express Charges	1.40	
		$ 1,524.05
Improvements, Construction and Acquisition—		
Land acquired for Park Purposes	$ 4,084.83	
Thieme Drive, Drains, Filling and Grading	725.98	
Lakeside Park and Parkstrip, Filling and Grading	104.84	

Wrecking Parham Warehouse	155.00	
Foster Park Fence, Material and Construct'n	404.43	
Foster Park Comfort Station and Tool House.	213.74	
Swinney Park Drive, Material and Construct'n	986.82	
Swinney Park Comfort Station	1,065.43	
Clinton St. Cement Walk along Lawton Park	502.50	
Pontiac Place Park Grading	153.00	
Hayden Park, Reconstructing Sidewalks	268.34	
Bowser Playgrounds	10.21	
Constructing Barn	640.09	
New Park Benches	167.50	
Basketball Banks	28.89	
Nursery Stock	109.05	
Trees and Shrubs	227.75	
Hot Bed Sash	20.35	
Lumber	29.48	
Crushed Stone	62.12	
Sand and Gravel	27.60	
Tennis Courts, Material	147.27	
Drinking Fountain Fixtures	16.12	
		$10.151.34

Forestry Department—

Salary of Forester	$ 725.04	
Pay Rolls Forestry Force	1,766.91	
Stationery and Supplies	88.45	
Tools and Equipment	187.99	
Repairs to Tools and Equipment	19.30	
Wagons and Equipment	73.15	
Freight and Drayage	43.66	
Express and Telegrams	1.33	
Motorcycle Accessories	24.35	
Motorcycle Repairs	15.60	
Motorcycle Maintenance	27.85	
Nursery Stock	298.88	
Guards and Stakes	53.06	
Cement and Tile	17.53	
Fertilizer	4.62	
Tree Surgery Materials	41.22	
Spraying Materials	7.25	
Sundries	11.00	
		$ 3,407.19

Park Maintenance—

Salary of Superintendent		$ 1,093.24
Pay Rolls, Maintenance	$12,188.18	
Pay Rolls, Cleaning Snow off Ice.	139.82	
		$12,328.00

Pay Rolls, Oiling Streets	36.00	
Pay Rolls, Playground Custodian	258.27	
Swinney Park Rent	600.00	
Lighting Parks and Lamp Renewals	128.46	
Freight and Drayage	43.65	
Fuel for Greenhouse	252.82	
Telephone Rental	36.00	
Fertilizer	153.75	
Road Oil	494.24	
Machinery and Tools	469.55	
Repairs to Machinery and Tools	144.61	
Repairs to Fountains	23.90	
Repairs to Bridges	51.42	
Repairs to Buildings	80.77	
Repairs to Hydrants	89.31	
Repairs to Flag Poles	15.00	
Repairs to Sewers	28.55	
Flower Pots	29.35	
Flower and Lawn Seeds	18.21	
Machine Oil	6.11	
Feed for Squirrels	5.43	
Sundries	15.75	
		$16,402.39
Total Disbursements		$40,457.07

OUR CITY'S SLOGAN
FORT WAYNE – WITH – MIGHT & MAIN
FORT WAYNE-1794

www.ingramcontent.com/pod-product-compliance
Lightning Source LLC
Chambersburg PA
CBHW060941050326
40689CB00012B/2531
9781013769030